Letts and **LONSDALE**

GCSE Success

Workbook

Science Foundation

Brian Arnold • Elaine Gill • Emma Poole

Contents

Physics

A balanced diet and nutrition

A

1 Which carbohydrate is found in milk? **(1 mark)**

a) glucose ☐
b) lactose ☐
c) maltose ☐
d) starch ☐

2 Glycerol is a component of which of the below? **(1 mark)**

a) carbohydrates ☐
b) fats ☐
c) proteins ☐
d) water ☐

3 Cholesterol is a type of **(1 mark)**

a) carbohydrate ☐
b) fat ☐
c) protein ☐
d) vitamin ☐

4 Which substance helps prevent constipation? **(1 mark)**

a) fibre ☐
b) minerals ☐
c) vitamins ☐
d) water ☐

5 What percentage of our body weight is water? **(1 mark)**

a) 55% ☐
b) 65% ☐
c) 75% ☐
d) 85% ☐

B

1 Complete the table. **(4 marks)**

Nutrient	Found in	Used for
carbohydrate	cereals	
fibre		moving food in gut
	all food and drink	cools us down
protein	lean meat	

2 State which two nutrients are missing from the table. Why does our body need them? **(4 marks)**

....

....

....

....

C

1 Food manufacturers are required to label their products with nutritional information.

Look at the two labels and answer the questions.

Brendan's Beans	
Typical values	**per 100 g**
Energy	400 kJ
Protein	7.0 g
Carbohydrate	20.0 g
Fats	1.0 g
of which saturated	0.5 g
Fibre	6.5 g
Salt	1.0 g

TOM's TOMATOES	
Typical values	**per 100 g**
Energy	70 kJ
Protein	1.0 g
Carbohydrate	3.0 g
Fats	0.1 g
of which saturated	trace
Fibre	0.5 g
Salt	trace

a) How much energy per 100 g, do you get from Brendan's beans? **(1 mark)**

....

....

....

b) Explain which food is better for those trying to lose weight. **(2 marks)**

....

....

....

....

c) Brendan's beans contain more energy. Explain one other reason why they are better for a teenager. **(1 mark)**

....

....

....

....

2 The table shows the energy needs of a female at different times in her life.

Life stage	Energy needs in a day in kJ
at infant school	8500
at secondary school	9500
adult	10000
pregnant adult	11500

a) Calculate the difference in her energy needs at infant and secondary school **(2 marks)**

....

....

b) Her activity is similar so why is there a difference? **(1 mark)**

....

....

c) Why does a pregnant female need more energy than a non pregnant one? **(1 mark)**

....

....

d) Name the disease commonly called 'the slimming disease'. **(1 mark)**

....

e) List two symptoms of the disease named in d. **(2 marks)**

....

....

....

....

The nervous system

A

1 The central nervous system (CNS) consists of **(1 mark)**

a) brain and the effectors ☐
b) brain and spinal cord ☐
c) spinal cord and receptors ☐
d) receptors and effectors ☐

2 Which structure is a sensory receptor? **(1 mark)**

a) liver ☐
b) kidney ☐
c) skin ☐
d) stomach ☐

3 Nerve cells are called **(1 mark)**

a) capillaries ☐
b) effectors ☐
c) neurones ☐
d) receptors ☐

4 Reflex actions help animals to **(1 mark)**

a) excrete ☐
b) grow ☐
c) reproduce ☐
d) survive ☐

5 The gap between two nerve cells is called a **(1 mark)**

a) fibre ☐
b) gland ☐
c) synapse ☐
d) vessel ☐

B

1 Complete the table by filling in the empty boxes. **(5 marks)**

Stimulus	Sense	Sense organ
	sight	eye
chemicals		taste buds (tongue)
sound waves	hearing	
pressure/temperature	touch	
	smell	nose

2 Complete the following passage. Use words from this list. **(6 marks)**

conscious   involuntary   learned
reflex   talking   voluntary

The actions you think about are called .... actions. They are under .... control.
They have to be .... e.g. ....
Actions that are automatic are called .... or .... actions.

C

1 a) The diagram shows a nerve cell. Name the parts labelled A, B and C. (3 marks)

b) How is this nerve cell different to a typical animal cell? (1 mark)

....

....

....

2 The passage of a nerve impulse always follows a set sequence of events. Use these words and phrases to show this sequence in the example given by filling in the boxes. (7 marks)

effector ⁶   motor neurone ⁵   receptor ²   relay neurone ⁴
response ⁷   sensory neurone ³   stimulus ¹

you sit on a drawing pin	
pain sensor in skin detects this	
nerve impulse sent to CNS	
nerve impulse transmitted in CNS	
nerve impulse sent to effector	
muscles contract	
you jump up	

The eye

A

1 **Binocular vision means** (1 mark)

a) one eye ☐
b) one eye on each side ☐
c) two eyes facing forward ☐
d) two eyes facing backward ☐

2 **Which two structures produce an image on the retina?** (1 mark)

a) conjunctiva and pupil ☐
b) cornea and lens ☐
c) iris and optic nerve ☐
d) sclera and brain ☐

3 **Which defect is inherited?** (1 mark)

a) blindness ☐
b) red-green colour blindness ☐

c) long sight ☐
d) short sight ☐

4 **In red-green colour blindness, which cells do not function correctly?** (1 mark)

a) blood ☐
b) cones ☐
c) nerve ☐
d) rods ☐

5 **In dim light, the pupil** (1 mark)

a) closes ☐
b) gets bigger ☐
c) gets smaller ☐
d) stays the same ☐

B

1 **Draw straight lines from the part of the eye to its description.** (7 marks)

cornea	helps focus the image
lens	a hole that allows light through (in front of the lens)
muscular iris	the protective, white outer layer of the eye
optic nerve	contains light sensitive cells
pupil	controls how much light enters the eye
retina	transparent window in the front of the eye
sclera	receives nerve impulses from the retina and sends them to the brain

2 **True or false?** (3 marks)

	true	false
a) Predators usually have monocular vision	☐	☐
b) Suspensory ligaments hold the lens in place	☐	☐
c) The eye is a sense organ	☐	☐

C

1 a) The diagram shows an eyeball and light rays from an object.

Name the structures labelled A, B and C.
(3 marks)

b) What defect does this eyeball show? Explain your answer. **(2 marks)**

....
....
....

2 a) The sentences describe how the eye sees things but they are in the wrong order.

Fill in the boxes to show the right order.
(4 marks)

A   The receptor cells in the retina send impulses to the brain.
B   Light from an object enters the eye through the cornea.
C   The brain interprets the image and you see the object the right way up.
D   The curved cornea and lens produce an image on the retina.
E   The image is upside down.

3 Complete the sentences using words from the list. **(3 marks)**

circular   radial   smaller

When a person moves from dim light into bright light, the iris reacts.

The .... muscles contract, the .... muscles relax.

The pupil gets .... Less light enters the eye.

4 The diagram shows a section through the eye.

Name the parts labelled **(6 marks)**

A ....
B ....
C ....
D ....
E ....
F ....

5 Describe the position of the eyes and the advantage this gives the animal in

a) binocular vision **(2 marks)**

....
....

b) monocular vision **(2 marks)**

....
....

The brain

A

1 The brain is protected by the   (1 mark)

a) heart ☐
b) ribs ☐
c) skull ☐
d) vertebrae ☐

2 Ecstasy affects the transmission of impulses across the   (1 mark)

a) cortex ☐
b) neurones ☐
c) medulla ☐
d) synapse ☐

3 Which disease has no cure?   (1 mark)

a) epilepsy ☐
b) grand mal ☐
c) meningitis ☐
d) petit mal ☐

4 How many neurones are there in the brain?   (1 mark)

a) hundreds ☐
b) thousands ☐
c) millions ☐
d) billions ☐

5 Ecstasy blocks the removal of which substance in the brain?   (1 mark)

a) haemoglobin ☐
b) plasma ☐
c) melanin ☐
d) seratonin ☐

B

1 The brain can suffer from many disorders.

a) Describe the symptoms of petit mal seizures.   (1 mark)

....

b) How are grand mal seizures different to petit mal seizures?   (1 mark)

....

c) Complete the table to show possible causes/increased risk and symptoms of the diseases shown.   (6 marks)

Disorder	Possible causes/increases risk	Symptoms/facts
strokes		
Parkinson's		
tumours		

C

1

a) The diagram shows a brain and the top of the spinal cord.

Add labelling lines and label

i) the cerebral cortex

ii) the medulla

iii) the cerebellum (3 marks)

b) Different parts of your brain are responsible for different functions.

Complete the table to show which part of the brain carries out each function (3 marks)

Function	Part of brain
controlling balance	
making decisions	
controlling heart and breathing rates	

c) On this diagram of the cerebral cortex, add labelling lines to show the areas that control the sensory areas, the motor areas, the memory and association areas. (3 marks)

d) Write down two other facts about the cerebral cortex. (2 marks)

....

....

....

....

2 Complete the sentences using words from the list. (6 marks)

electrical impulses   environment   learn
muscles   neurons   pathways

The brain works by sending .... received from the sense organs to the .... In mammals, the brain is complex and involves billions of .... that allow learning by experience and behaviour.
The interaction between mammals and their .... results in nerve pathways forming in the brain. When mammals learn from experience, ... in the brain become more likely to transmit impulses than others, which is why it is easier to .... through repetition.

Causes of disease

A

1 Which type of microbe reproduces by producing spores? *(1 mark)*

a) all of them ☐
b) bacteria ☐
c) fungi ☐
d) viruses ☐

2 Which disease is caused by a bacterium? *(1 mark)*

a) athletes foot ☐
b) HIV ☐
c) influenza ☐
d) tuberculosis/T.B. ☐

3 Which disease is caused by a virus? *(1 mark)*

a) common cold ☐
b) food poisoning ☐
c) whooping cough ☐
d) ringworm ☐

4 Which type of microbe is used in bread making? *(1 mark)*

a) all of them ☐
b) bacteria ☐
c) fungi ☐
d) viruses ☐

5 Which type of organism is found when decomposition occurs? *(1 mark)*

a) all of them ☐
b) insects ☐
c) fungi ☐
d) viruses ☐

B

1 The table shows features of bacteria and viruses. Place a tick (if feature present) or a cross (if feature absent) in each box. *(4 marks)*

Feature	Bacteria	Viruses
cell wall		
protein coat		
respire, feed and move		
reproduce inside living cells		

2 Name two structures found in plant cells that are not present in bacteria or viruses. *(2 marks)*

....

3 List three symptoms of infection. *(3 marks)*

....

....

C

1 a) Tuberculosis destroys lung tissue.

How is it spread? (2 marks)

....

....

....

b) List three things that helped to reduce tuberculosis. (3 marks)

....

....

....

2 Some diseases are not caused by microbes.
Complete the table to show what causes the diseases listed. (4 marks)

Disease	Cause
anaemia	
cancer	
red-green colour blindness	
scurvy	

3 Name three ways in which microbes can enter the body.

For each give an example of a disease spread in this way. (6 marks)

....

....

....

....

....

....

....

Defence against disease

A

1 Which organ produces acid to kill bacteria? **(1 mark)**

a) heart ☐
b) liver ☐
c) kidney ☐
d) stomach ☐

2 Which is the resistant bacterium found in hospitals? **(1 mark)**

a) AIDS ☐
b) HIV ☐
c) MRSA ☐
d) STD ☐

3 Lymphocytes produce chemicals called **(1 mark)**

a) antibiotics ☐
b) antibodies ☐
c) antigens ☐
d) antiseptics ☐

4 Which microbe cannot be grown on nutrient agar? **(1 mark)**

a) all of them ☐
b) bacteria ☐
c) fungi ☐
d) viruses ☐

5 In the MMR vaccine, R stands for **(1 mark)**

a) german measles/rubella ☐
b) measles ☐
c) meningitis ☐
d) rabies ☐

B

1 a) Antibiotics are used to kill bacteria inside the body.

Which microbe cannot be killed by antibiotics? **(1 mark)**

....

b) Why do scientists have to find new antibiotics to kill bacteria? **(2 marks)**

....

....

2 Complete the passage using words from the list. **(4 marks)**

antitoxins   lymphocytes   phagocytes   white blood cells

If microbes get into your body, ... travelling around in your blood spring into action. White blood cells can make chemicals called .... that destroy the toxins produced by bacteria. White blood cells called .... try to engulf bacteria or viruses before they have a chance to do any harm. If the microbes are in large numbers, however, the other type of white blood cell, called ...., are involved.

C

1 a) Explain how your skin prevents microbes from entering your body. (3 marks)

....

....

....

b) When you cut yourself, microbes could get into your blood.

How does the blood kill microbes? (2 marks)

....

....

....

2 a) If you come into contact with a person suffering from mumps, the microbes could enter your body. Explain how you will develop a natural immunity to mumps. (5 marks)

....

....

....

....

....

b) Immunity can be also be artificial.

How is active artificial immunity different to natural immunity? (2 marks)

....

....

....

c) How is passive artificial immunity different to active artificial immunity? (1 mark)

....

....

....

Drugs

A

1 Which drug has recently had its classification changed? **(1 mark)**

a) cannabis ☐
b) cocaine ☐
c) heroin ☐
d) paracetamol ☐

2 Which drug makes you see and hear things that do not exist? **(1 mark)**

a) depressants ☐
b) hallucinogens ☐
c) painkillers ☐
d) stimulants ☐

3 Which drug is not a hallucinogenic drug? **(1 mark)**

a) cannabis ☐
b) cocaine ☐
c) ecstasy ☐
d) LSD ☐

4 Which drug is addictive and increases the risk of HIV? **(1 mark)**

a) cannabis ☐
b) heroin ☐
c) paracetemol ☐
d) penicillin ☐

5 Tranquilisers and sleeping pills are examples of which type of drug? **(1 mark)**

a) depressants ☐
b) hallucinogens ☐
c) sedatives ☐
d) stimulants ☐

B

1 Complete the sentences.
Use words from this list. **(5 marks)**

behaviour   brain   chemicals   nervous system   useful

Drugs are powerful ....; they alter the way the body works, often without you

realising it. There are .... drugs such as antibiotics like penicillin, but these

can be dangerous if misused. Drugs affect the .... and ....,

which in turn affects .... and risk of infection.

2 True or false? **(3 marks)**

	true	false
a) Alcohol increases the activity of the brain	☐	☐
b) Alcohol is a depressant	☐	☐
c) Alcohol can cause cirrhosis	☐	☐

C

1 The information is about birth weights of babies.

Mother	Average birth weight in kgs
non-smoker	3.50
smoker	2.90

a) How does smoking affect the birth weight of babies? **(1 mark)**

....

....

....

....

....

b) Explain how smoking causes this effect. **(2 marks)**

....

....

....

....

2 Complete the table by filling in the organs each drug affects. **(6 marks)**

Drug	Organs
alcohol	1..
	2..
solvents	1..
	2..
	3..
painkillers	1..

3 Name two diseases which may be caused by smoking. **(2 marks)**

....

....

....

4 a) What is the addictive substance in cigarette smoke? **(1 mark)**

....

....

....

b) Name the two other harmful chemicals in cigarette smoke and explain why they are harmful. **(4 marks)**

....

....

....

....

....

....

Homeostasis and diabetes

A

1 Some diabetics can control the disease by a diet low in **(1 mark)**

a) fat ☐
b) glucose ☐
c) protein ☐
d) starch ☐

2 Which scientists discovered insulin? **(1 mark)**

a) Banting and Best ☐
b) Jenner ☐
c) Darwin ☐
d) Watson and Crick ☐

3 Which substance is mainly excreted by the kidneys? **(1 mark)**

a) carbon dioxide ☐
b) salts ☐
c) urea ☐
d) water ☐

4 Which organ controls water loss in the body? **(1 mark)**

a) heart ☐
b) liver ☐
c) kidneys ☐
d) pancreas ☐

5 Which is a symptom of diabetes? **(1 mark)**

a) breathlessness ☐
b) cold ☐
c) hot ☐
d) thirst ☐

B

1 Which organ loses water in the urine? **(1 mark)**

....

2 Homeostasis is the mechanism by which the body maintains a constant internal environment. Describe three things the body keeps constant. **(3 marks)**

....

....

3 On the diagram which letter shows **(4 marks)**

i)   the renal artery
ii)   the ureter
iii) where urine is made
iv) where urine is stored

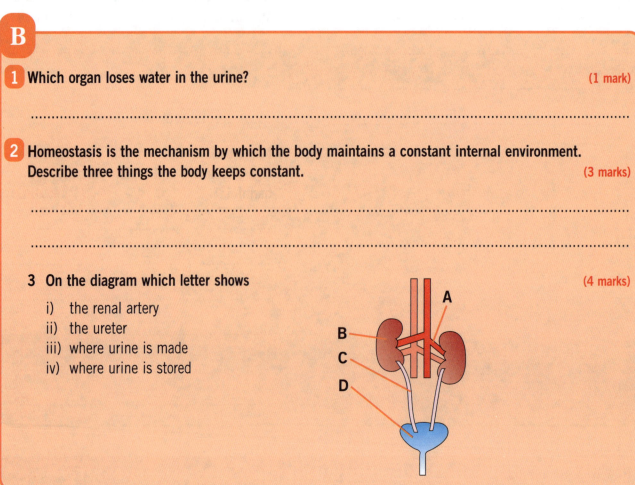

C

1 **Complete the sentences using words from the list.** (9 marks)

glucagon   glucose   glycogen   high   hormones   insulin   liver   low   normal

The pancreas is an organ involved in homeostasis; it maintains the level of .... (sugar) in the blood so that there is enough for respiration. The pancreas secretes two .... into the blood, insulin and glucagon. If blood sugar levels are too .... , which could be the case after a high carbohydrate meal, special cells in the pancreas detect these changes and release .... The .... responds to the amount of insulin in the blood and takes up glucose and stores it as .... Blood sugar levels return to ....

If blood sugar levels are too .... , which could be the case during exercise, the pancreas secretes .... This stimulates the conversion of stored glycogen in the liver back into glucose which is released into the blood. Blood sugar levels return to normal.

2 **Three samples of liquid were taken, one from the blood entering the kidneys, one from the kidneys and one from the bladder. The samples were analysed to find the percentage of glucose, protein, urea and water present.**

The results are shown in the table.

Substance	Percentage of substance in		
	Blood	Filtrate in kidneys	Urine in bladder
glucose	0.1	0.1	0.0
protein	9.0	0.0	0.0
urea	0.03	0.03	2.0
water	90.0	99.0	97.0

a) **Which substance is present in the blood but not in the filtrate?** (1 mark)

....

b) **Which substance is more concentrated in the urine than the filtrate?** (1 mark)

....

c) **There is no glucose in the urine, but it is present in the filtrate.**
   Explain this difference. (2 marks)

....

....

....

How well did you do?   ✗ **0-10** Try again   **11-16** Getting there   **17-21** Good work   **22-26** Excellent! ✓

The menstrual cycle

A

1 **Puberty is the first stage of** (1 mark)

a) adolescence ☐
b) childhood ☐
c) infancy ☐
d) middle age ☐

2 **The egg is released during which days in a typical menstrual cycle?** (1 mark)

a) 1–5 ☐
b) 5–14 ☐
c) 14–28 ☐
d) 28–5 ☐

3 **A typical menstrual cycle lasts how many days?** (1 mark)

a) 7 ☐
b) 14 ☐
c) 21 ☐
d) 28 ☐

4 **The male sex hormone is called** (1 mark)

a) insulin ☐
b) oestrogen ☐
c) progesterone ☐
d) testosterone ☐

5 **Which hormone can be used as an oral contraceptive?** (1 mark)

a) FSH ☐
b) oestrogen ☐
c) progesterone ☐
d) testosterone ☐

B

1 Changes occur in the human body during puberty.
Some occur in boys and girls, whilst others only occur in one of the sexes.
Place a tick in the box to show which change happens to which sex. (8 marks)

Change	Boys	Girls
breasts develop		
genitals develop		
hair grows under the arms		
hair grows on the face and body		
menstruation begins		
pubic hair grows		
sperm production begins		
voice deepens		

2 What are the two main jobs of the menstrual cycle? (2 marks)

....

....

C

1 The diagram represents a typical menstrual cycle.

A

days 1–5

B

days 14–28

C

days 5–14

a) Which letter represents

   i) ovulation.................................

   ii) uterus ready for implantation

   ....

   iii) menstruation......................... **(3 marks)**

b) Explain what happens during days 1–5.

(2 marks)

....

....

....

....

c) The ovaries release oestrogen. What does it do? **(3 marks)**

....

....

....

....

2 a) FSH is called a 'fertility drug' Explain why. **(2 marks)**

....

....

....

....

b) What is a disadvantage of giving a woman FSH? **(1 mark)**

....

....

....

....

3 Progesterone is also produced during the menstrual cycle. What does it do and what happens when production stops? **(2 marks)**

....

....

....

....

4 IVF stands for in vitro fertilisation. What does it involve? **(3 marks)**

....

....

....

....

5 How does oestrogen work as an oral contraceptive? **(2 marks)**

....

....

How well did you do? ✗ 0-13 Try again   14-19 Getting there   20-26 Good work   27-33 Excellent! ✓

Genetics and variation

A

1 How many genes are there in a typical human cell? (1 mark)

- a) tens ☐
- b) hundreds ☐
- c) thousands ☐
- d) millions ☐

2 Who will have the same genes? (1 mark)

- a) mother and son ☐
- b) brother and sister ☐
- c) identical twins ☐
- d) non-identical twins ☐

3 Which of these is is controlled by the environment? (1 mark)

- a) blood group ☐
- b) eye colour ☐
- c) gender ☐
- d) height ☐

4 Which of these is not controlled by genes only? (1 mark)

- a) shoe size ☐
- b) ear lobes ☐
- c) IQ ☐
- d) weight ☐

5 Which of these factors is inherited? (1 mark)

- a) finger length ☐
- b) playing the piano ☐
- c) scars ☐
- d) speaking Welsh ☐

B

1 Identical twins were separated at birth and brought up in different environments.

a) Name two features that would be identical in the twins. (2 marks)

....

b) Name two features that could be different. (2 marks)

....

2 True or false? (3 marks)

	true	false
a) Living things that belong to the same species are all slightly different.	☐	☐
b) Variation can be between species or within species.	☐	☐
c) Genetics or the environment determines how we look and behave.	☐	☐

C

1 A gardener wanted to grow more geraniums. He collected the seeds of his plants and grew them. He also cut off growing tips of healthy plants and placed them in compost and they grew.

a) State three environmental factors that can affect seed germination and plant growth.
(3 marks)

....

....

....

....

....

....

b) Describe and explain the flower colour of the plants grown from cuttings.   (2 marks)

....

....

....

....

....

....

c) What name is given to the plants grown from cuttings?   (1 mark)

....

....

....

....

....

....

2 Describe how you could show if differences in the colour of geraniums was due to the environment.   (5 marks)

....

....

....

....

....

....

....

....

....

....

....

....

3 We all look different because of the way our genes are inherited and the environment.

Name four features that are due entirely to our genetics.   (4 marks)

....

....

....

....

4 Plants are affected by environmental factors.

Name four environmental factors that could affect plants.   (4 marks)

....

....

....

....

Genetics

A

1 Which monk discovered the principles behind genetics? **(1 mark)**

a) Jenner ☐
b) Mendel ☐
c) Newton ☐
d) Pasteur ☐

2 Which disease can be inherited? **(1 mark)**

a) chickenpox ☐
b) common cold ☐
c) cystic fibrosis ☐
d) german measles ☐

3 What do genes code for? **(1 mark)**

a) acids ☐
b) chromosomes ☐
c) characteristics ☐
d) proteins ☐

4 Which human problem could be treated using gene therapy in the future? **(1 mark)**

a) measles ☐
b) poor growth ☐
c) short sight ☐
d) deafness ☐

5 Which of the following organisms are used to produce human insulin in genetic engineering? **(1 mark)**

a) bacteria ☐
b) fungi ☐
c) viruses ☐
d) rabbits ☐

B

1 What do the words mean?
Fill in the boxes. Use words from this list. **(6 marks)**

dominant   genotype   heterozygous   homozygous   phenotype   recessive

Definition	Word
different alleles	
the stronger allele	
the type of alleles the organism has	
the weaker allele	
what the organism looks like	
both alleles the same	

2 Complete the sentences. Use words from this list. **(4 marks)**

chromosomes   disease   insulin   protein

Most cells contain a nucleus. Inside the nucleus there are thread-like structures called .... On each of these thread-like structures are thousands of codes. Each code is a gene. Each gene determines what .... is made.

When the code is wrong a person may suffer from a genetic .... Genes can be manipulated to help overcome this. The gene that codes for .... can be found in human pancreas cells.

C

1 Peas can be round or wrinkled.

R represents the dominant allele, round and r represents the recessive allele, wrinkled.

a) Complete the punnet square.

(4 marks)

	r	r
R		
r		

b) What type of pea would these genotypes code for? (4 marks)

i) RR ....

ii) Rr ....

iii) rR ....

iv) rr ....

2 a) A child who suffers from cystic fibrosis has a faulty gene. They could be helped by gene therapy. What is gene therapy?.........(1 mark)

....

....

....

b) Describe what the doctors would try to do. (1 mark)

....

....

....

....

c) What problems might they encounter? (2 marks)

....

....

....

....

....

3 a) Which industries benefit from genetic engineering? (2 marks)

....

....

....

....

....

b) Name three examples where genetic engineering is useful. (3 marks)

....

....

....

....

....

Inherited diseases

A

1 Uncontrolled jerky movements are a symptom of which disease? **(1 mark)**

a) chickenpox ☐
b) cystic fibrosis ☐
c) Huntington's disease ☐
d) german measles ☐

2 Which disease affects the brain? **(1 mark)**

a) measles ☐
b) Huntington's disease ☐
c) leukaemia ☐
d) sickle cell anaemia ☐

3 Name the cell that has not yet specialised. **(1 mark)**

a) blood ☐
b) nerve ☐
c) stem ☐
d) sperm ☐

4 Which of these diseases is an hereditary disease? **(1 mark)**

a) bronchitis ☐
b) cancer ☐
c) Huntington's disease ☐
d) obesity ☐

5 What is an inherited disease? **(1 mark)**

a) disease caused by microbes ☐
b) disease passed from person to person ☐
c) disease passed on from parent to child genetically ☐
d) self-inflicted disease ☐

B

1 Huntington's disease affects one in 20 000 people, so is a rare disease.
Huntington's disease is caused by a dominant allele (H).
The diagram shows a family tree.

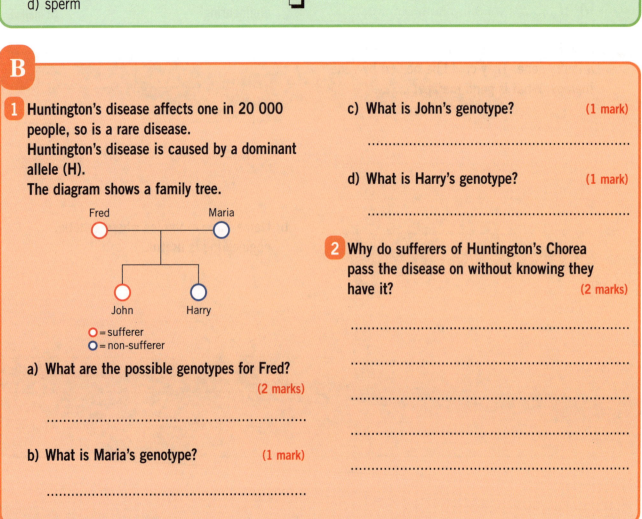

Fred   Maria

John   Harry

○ = sufferer
○ = non-sufferer

a) What are the possible genotypes for Fred? **(2 marks)**

....

b) What is Maria's genotype? **(1 mark)**

....

c) What is John's genotype? **(1 mark)**

....

d) What is Harry's genotype? **(1 mark)**

....

2 Why do sufferers of Huntington's Chorea pass the disease on without knowing they have it? **(2 marks)**

....
....
....
....
....

C

1 Cystic fibrosis is caused by a recessive allele, carried by about one person in 20.

The genotype of a sufferer is ff.

a) What are the possible genotypes of a non-sufferer? **(2 marks)**

....

....

....

b) What is the genotype of a carrier? **(1 mark)**

....

....

....

c) What is a carrier? **(3 marks)**

....

....

....

....

d) What is the probability of two carriers having a child that suffers? **(1 mark)**

....

....

....

e) What is the probability of two carriers having a child that is a carrier? **(1 mark)**

....

....

....

2 a) What are the symptoms of cystic fibrosis? **(4 marks)**

....

....

....

....

....

....

....

b) What treatment is available? **(2 marks)**

....

....

c) What does a genetic counsellor do? **(2 marks)**

....

....

3 a) Where are stem cells found? **(3 marks)**

....

....

b) What potential do stem cells have in treating inherited disease? **(2 marks)**

....

....

c) Stem cell research is in its infancy. What is needed to expand the research programme? **(3 marks)**

....

....

How well did you do? ✗ **0-14** Try again   **15-21** Getting there   **22-28** Good work   **29-36** Excellent! ✓

Selective breeding

A

1 Cows are selectively bred to (1 mark)

a) have bigger feet ☐
b) produce less meat ☐
c) produce more milk ☐
d) have stronger legs ☐

2 Which of the following produces a clone? (1 mark)

a) asexual reproduction ☐
b) fertilisation ☐
c) sexual reproduction ☐
d) selective breeding ☐

3 What is a clone? (1 mark)

a) animals that live together ☐
b) genetically identical organisms ☐
c) organisms that are similar ☐
d) plants with no flowers ☐

4 Taking sperm from the best bull and putting it into the best cow is called (1 mark)

a) artificial insemination ☐
b) artificial selection ☐
c) natural selection ☐
d) variation ☐

5 Selective breeding is another name for (1 mark)

a) artificial selection ☐
b) evolution ☐
c) natural selection ☐
d) variation ☐

B

1 If animals or plants are continually bred from the same best animals or plants, the animals and plants will all be very similar.

a) What are the disadvantages of this? (2 marks)

....

b) Why is important to keep wild varieties/rare breeds alive? (1 mark)

....

2 Embryo transplants are used in the breeding of cattle.

a) Write a letter in each box to explain what is happening. (4 marks)

    A eggs taken from best cows
    B embryos implanted into surrogate cows
    C sperm taken from best bull
    D fertilised egg grows into embryo

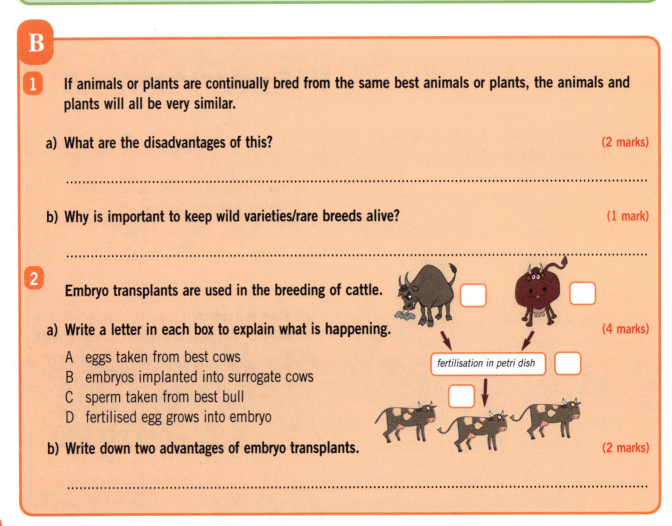

fertilisation in petri dish

b) Write down two advantages of embryo transplants. (2 marks)

....

1 The sentences describe the stages involved in artificial selection.

They are in the wrong order.

Using the letters, write the correct order in the boxes. **(5 marks)**

P breed them together using sexual reproduction.
Q repeated over generations
R all offspring have the desired characteristics.
S select the individuals with the best characteristics.
T the best offspring are selected and are bred together.

2 A grower had some scented roses.

He took some cuttings to produce more scented roses.

a) What type of reproduction is this? **(1 mark)**

....
....
....
....

b) Why did he use this type of reproduction? **(1 mark)**

....
....
....
....

3 a) Strawberries are bred for large tasty berries. Suggest why. **(1 mark)**

....
....
....

b) Once a large red strawberry has been grown, the plants can be grown by tissue culture.

Stage 1   *Stage 2*

growth medium

Stage 3

tasty strawberry

The diagrams show some of the stages in the process.

Explain what is happening at each stage. **(3 marks)**

....
....
....
....

c) What are the advantages of growing plants by tissue culture? **(3 marks)**

....
....
....
....

Pyramids

A

1 **What does a pyramid of numbers tell us?** (1 mark)

- a) animals that live together ☐
- b) number of organism in a food chain ☐
- c) organisms that are similar ☐
- d) animals that breed together ☐

2 **Why do food chains rarely have more than four or five links in them?** (1 mark)

- a) animals at the beginning are too big ☐
- b) energy is lost ☐
- c) producers cannot make enough food ☐
- d) too many plants die ☐

3 **What is the name for a level in a food chain?** (1 mark)

- a) first level ☐
- b) floor level ☐
- c) plant level ☐
- d) trophic level ☐

4 **What does a pyramid of biomass tell us?** (1 mark)

- a) animals that live together ☐
- b) mass of organism in a food chain ☐
- c) organisms that are different ☐
- d) organisms that are similar ☐

5 **How can we improve the efficiency of food production?** (1 mark)

- a) grow less food ☐
- b) increase the number of links in a food chain ☐
- c) grow more food ☐
- d) reduce number of links in food chain ☐

B

1 **Here is a food chain.**

grass → rabbit → owl

a) What does the rabbit eat? (1 mark)

....

In this chain there were 1000 grass plants, 10 rabbits and 1 owl.

b) Draw a pyramid of numbers for this food chain. (3 marks)

Another chain had 1 rose bush, 1000 greenfly and 10 ladybirds

c) Draw a pyramid of numbers for this food chain. (3 marks)

d) Why is a pyramid of numbers misleading? (1 mark)

....

C

1 Here is a pyramid of numbers.

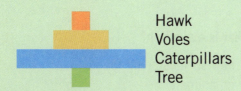

Hawk
Voles
Caterpillars
Tree

a) Draw a pyramid of biomass for this food chain. **(4 marks)**

b) Explain how you calculate the biomass from a pyramid of numbers. **(2 marks)**

....

....

....

....

....

....

....

....

....

2 The diagram shows a sheep and its energy gains and losses.

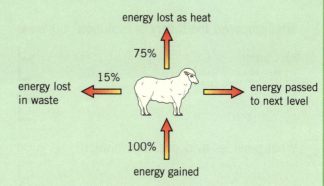

energy lost as heat

75%

15%

energy lost in waste

energy passed to next level

100%

energy gained

a) Calculate how much energy is passed on to the next level. **(2 marks)**

....

....

....

b) How is most energy lost from this sheep? **(1 mark)**

....

....

....

c) Suggest what form the waste might take. **(2 marks)**

....

....

....

d) How is energy gained? **(1 mark)**

....

....

Evolution

A

1 Who proposed the theory of evolution? (1 mark)

- a) Darwin ☐
- b) Newton ☐
- c) Pasteur ☐
- d) Watson ☐

2 What provides evidence for evolution? (1 mark)

- a) animals ☐
- b) fossils ☐
- c) plants ☐
- d) viruses ☐

3 What is another name for survival of the fittest? (1 mark)

- a) artificial selection ☐
- b) natural selection ☐
- c) sexual reproduction ☐
- d) variation ☐

4 What is extinction? (1 mark)

- a) new species being found ☐
- b) no animals ☐
- c) species dying out ☐
- d) species hiding ☐

5 What is a fossil? (1 mark)

- a) bones ☐
- b) insects ☐
- c) minerals ☐
- d) remains of dead organisms ☐

B

1 Name three factors that might prevent offspring surviving. (3 marks)

....

2 Darwin visited the Galapagos Islands.

a) Write down the four observations that he made which formed the basis of the theory of evolution. (4 marks)

....
....
....
....

b) What did Darwin conclude from these observations? (3 marks)

....
....
....

C

1 The peppered moth is an example of evolution.

There are two forms, one is dark in colour, the other is light-coloured.

They live in woodlands on lichen covered trees.

The light coloured moth was common before the industrial revolution.

a) Suggest how the dark coloured moth originated. **(1 mark)**

....

....

....

....

....

b) Why did the light coloured moth decline during the industrial revolution? **(2 marks)**

....

....

....

....

....

....

....

c) Why did the darker moth survive during the industrial revolution? **(1 mark)**

....

....

....

....

....

d) Now we have smoke free areas, what is happening to the numbers of dark and light coloured moths? **(2 marks)**

....

....

....

....

....

2 Explain how fossils are formed. **(4 marks)**

....

....

....

....

....

....

....

....

....

....

Adaptation and competition

A

1 **What is a limiting factor?** (1 mark)

a) something that increases the size of a population ☐

b) something that changes the gene pool ☐

c) something that stops a population becoming too large ☐

d) something that increases oxygen ☐

2 **What does a camel store in its hump?** (1 mark)

a) fat ☐

b) glucose ☐

c) protein ☐

d) water ☐

3 **Which is not a feature of the polar bear?** (1 mark)

a) good swimmer ☐

b) coloured coat ☐

c) sharp claws ☐

d) greasy fur ☐

4 **What do plants compete for that animals do not?** (1 mark)

a) light ☐

b) nutrients ☐

c) space ☐

d) water ☐

5 **Which feature would you find in a predator?** (1 mark)

a) all round vision ☐

b) brightly coloured ☐

c) poor hearing ☐

d) sharp claws ☐

B

1 **Match these words to their definitions. Write your answers in the boxes.** (4 marks)

community   ecosystem   habitat   population

Definition	Word
where organism lives	
all one type of animal or plant	
living things in the habitat	
all the living things and their physical environment	

2 **Name four factors that limit a population** (4 marks)

....

....

3 **a) What is a predator?** (1 mark)

....

b) What is prey? (1 mark)

....

C

1 Look at the picture of a polar bear.

Polar bears live in the arctic where it is very cold. They swim in freezing water.

Explain what features they have which adapt them to this environment. **(5 marks)**

....
....
....
....
....
....
....
....
....
....
....
....
....

2 The diagram shows the relationship between predator and prey.

a) Describe the changes in prey population.
(1 mark)

....
....

b) Describe the changes in predator population. **(1 mark)**

....
....

c) i) When does prey population decrease?
(1 mark)

....

ii) Explain why. **(1 mark)**

....

d) i) What causes the prey population to increase? **(1 mark)**

....

ii) Explain why. **(1 mark)**

....

How well did you do? 0-10 Try again   11-16 Getting there   17-21 Good work   22-26 Excellent!

35

Environmental damage 1

A

1 Which of the following is caused by deforestation? **(1 mark)**

a) increase in habitats ☐
b) increases the amount of carbon dioxide in the air ☐
c) increase in rainfall ☐
d) increase in soil erosion ☐

2 Burning fossil fuel can lead to an increase in what? **(1 mark)**

a) deforestation ☐
b) eutrophication ☐
c) greenhouse effect ☐
d) organic farming ☐

3 What is used to help plants grow in organic farming? **(1 mark)**

a) minerals ☐
b) manure ☐
c) NPK fertiliser ☐
d) pesticides ☐

4 What is needed to produce more food in intensive farming? **(1 mark)**

a) fertilisers ☐
b) land ☐
c) water ☐
d) recycling ☐

5 Which is an not an alternative energy source? **(1 mark)**

a) wave power ☐
b) fossil fuel ☐
c) solar power ☐
d) wind energy ☐

B

1 Complete the sentences. Use words from the list. **(5 marks)**

fertilisers   food   intensive   minerals   pesticides

Farming has had to become more .... to try and provide more ....
from a given area of land. Many people regard this type of farming of animals as cruel.
In order to produce more food from the land, ... and .... are
needed. Chemicals called .... are used to kill pests.

2 Explain how deforestation causes problems for the environment. **(5 marks)**

....

....

....

....

....

C

1 Look at the diagram.

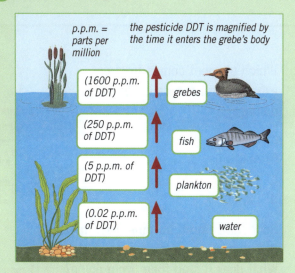

p.p.m. = parts per million

the pesticide DDT is magnified by the time it enters the grebe's body

(1600 p.p.m. of DDT) — grebes

(250 p.p.m. of DDT) — fish

(5 p.p.m. of DDT) — plankton

(0.02 p.p.m. of DDT) — water

a) Explain how the pesticide gets into the food chain.　**(3 marks)**

....

....

....

....

....

b) How much DDT is in the plankton?　**(1 mark)**

....

....

....

c) Why does the grebe die?　**(4 marks)**

....

....

....

....

....

2 A tomato grower wanted to kill the red spider mites which were eating his tomato plants.

He introduced another predatory mite into his greenhouses.

a) How would this help the grower to get rid of the red spider mites?　**(1 mark)**

....

....

....

....

b) What is this type of control called?　**(1 mark)**

....

....

....

....

c) What are the advantages of this sort of control?　**(2 marks)**

....

....

....

....

d) What are the disadvantages of this sort of control?　**(1 mark)**

....

....

....

How well did you do?　✗ **0-11** Try again　**12-16** Getting there　**17-22** Good work　**23-28** Excellent! ✓

Environmental damage 2

A

1 Which gas contributes to the greenhouse effect? **(1 mark)**

a) chlorine ☐
b) methane ☐
c) nitrogen ☐
d) oxygen ☐

2 What is a pollutant? **(1 mark)**

a) substance that is useful to living things ☐
b) substance that comes from the earth ☐
c) substance that is harmful to living things ☐
d) substance that is made by animals ☐

3 What is sustainable development? **(1 mark)**

a) keeping things the same ☐
b) looking after the earth for the future ☐
c) monitoring resources ☐
d) planting more trees ☐

4 Which element comes from car exhausts and can cause brain damage? **(1 mark)**

a) chlorine ☐
b) lead ☐
c) nitrogen ☐
d) oxygen ☐

5 Soot comes from burning fossil fuels. It is deposited on the leaves of plants. What does it stop them from doing? **(1 mark)**

a) breathing ☐
b) excreting ☐
c) photosynthesising ☐
d) reproducing ☐

B

1 True or false? **(4 marks)**

	true	false
a) burning fossil fuels is a cause of acid rain	☐	☐
b) catalytic converters in cars increase emissions of harmful gases	☐	☐
c) using unleaded petrol reduces the greenhouse effect	☐	☐
d) carbon dioxide is a greenhouse gas	☐	☐

2 a) What does CFC stand for? **(1 mark)**

....

b) Where do they come from? **(1 mark)**

....

c) What effect do they have in the atmosphere? **(2 marks)**

....

d) What consequence can this have for humans? **(1 mark)**

....

C

1 Look at the diagram. It shows some sources of pollutants of the air.

power station

a) List three gases which pollute the air. **(3 marks)**

....

....

....

b) How do these gases contribute to acid rain? **(3 marks)**

....

....

....

....

c) What effect does acid rain have

i) on buildings **(1 mark)**

....

ii) on lakes **(2 marks)**

....

....

iii) on trees **(1 mark)**

....

2 Evidence is being collected that shows that the world is warming up.

a) What causes this? **(1 mark)**

....

....

....

....

b) What is this warming up called? **(1 mark)**

....

....

....

....

c) Look at the diagram.

i) .... amount of carbon dioxide in atmosphere

ii) .... carbon dioxide in atmosphere

iv) .... are reflected

iii) .... rays from Sun

v) .... rays from sun

vi) .... are reflected so atmosphere warms up

Earth 1950

Earth 2002

Explain this warming up by using these words to fill in the spaces in the diagram. A word can be used more than once **(6 marks)**

small   more   most   less   UV

How well did you do? 0-13 Try again 14-19 Getting there 20-26 Good work 27-32 Excellent!

Ecology and classification

A

1 **What does an ecologist study?** (1 mark)

a) effect of the sun ☐
b) living things in their habitat ☐
c) resources needed by humans ☐
d) results of planting more trees ☐

2 **Who devised a system of classification?** (1 mark)

a) Banting ☐
b) Best ☐
c) Linnaeus ☐
d) Darwin ☐

3 **How many kingdoms are there now?** (1 mark)

a) 2 ☐
b) 3 ☐
c) 4 ☐
d) 5 ☐

4 **Which is an abiotic factor?** (1 mark)

a) colour ☐
b) length of beak ☐
c) light intensity ☐
d) number of legs ☐

5 **What bird did Darwin study on the Galapagos Islands?** (1 mark)

a) blackbirds ☐
b) finches ☐
c) owls ☐
d) eagles ☐

B

1 **a)** When organisms are put into groups, there is an order of classification.

Here are the group names but in the wrong order.

Rewrite them in the correct order. (7 marks)

Class   Family   Genus   Kingdom   Order   Phylum   Species

....

....

b) What is a species? (2 marks)

....

....

2 **Name the five groups of vertebrates.** (5 marks)

....

....

C

1 Use this key to identify these four organisms.

(4 marks)

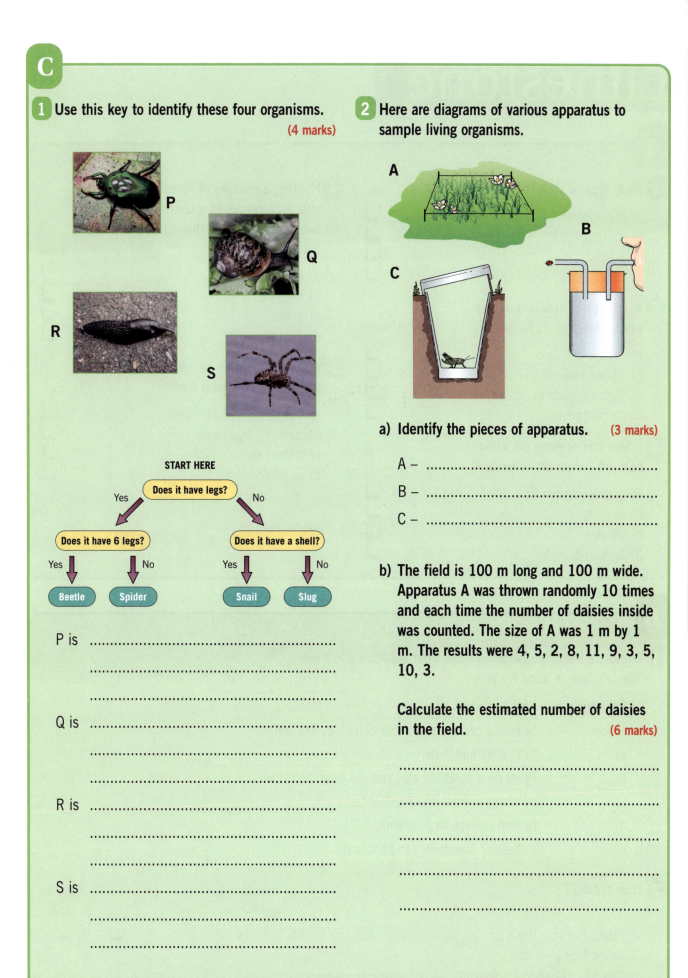

START HERE

Does it have legs?

Yes   No

Does it have 6 legs?   Does it have a shell?

Yes   No   Yes   No

Beetle   Spider   Snail   Slug

P is ....
....
....

Q is ....
....
....

R is ....
....
....

S is ....
....
....

2 Here are diagrams of various apparatus to sample living organisms.

A

B

C

a) Identify the pieces of apparatus. (3 marks)

A – ....

B – ....

C – ....

b) The field is 100 m long and 100 m wide. Apparatus A was thrown randomly 10 times and each time the number of daisies inside was counted. The size of A was 1 m by 1 m. The results were 4, 5, 2, 8, 11, 9, 3, 5, 10, 3.

Calculate the estimated number of daisies in the field. (6 marks)

....
....
....
....
....

How well did you do? ✗ **0-12** Try again   **13-19** Getting there   **20-25** Good work   **26-32** Excellent! ✓

Limestone

A

1 What type of rock is limestone?   (1 mark)

a) sedimentary
b) metamorphic
c) igneous
d) mineral

2 What type of reaction occurs when limestone is heated?   (1 mark)

a) exothermic
b) neutralisation
c) thermal decomposition
d) displacement

3 Which gas is given off when calcium carbonate is heated?   (1 mark)

a) carbon monoxide
b) oxygen
c) nitrogen
d) carbon dioxide

4 Heating a mixture of limestone, sand and soda produces a useful new substance. What is this substance called?   (1 mark)

a) concrete
b) cement
c) glass
d) quicklime

5 What is the name of the chemical compound in slaked lime?   (1 mark)

a) calcium carbonate
b) calcium hydroxide
c) limestone
d) calcium oxide

B

**1 Complete the table below to show what substance is made.
The first one is done for you.**   (5 marks)

Substance	How it is made
concrete	by mixing cement, sand, rock chippings and water.
a)	by heating limestone.
b)	by roasting powdered clay and powdered limestone.
c)	by mixing cement, sand and water.
d)	by adding water to quicklime.
e)	by heating limestone, sand and soda.

2 True or false?   (5 marks)

	true	false
a) Powdered limestone can be used to neutralise the acidity in lakes caused by acid rain.	☐	☐
b) The main chemical compound in limestone is calcium oxide.	☐	☐
c) Concrete is strong but very expensive.	☐	☐
d) Slaked lime will neutralise the acidity in lakes faster than powdered limestone.	☐	☐
e) When calcium oxide is reacted with water, the product is calcium carbonate.	☐	☐

C

1 When sodium hydrogen carbonate is heated it reacts to form sodium carbonate, carbon dioxide and water.

a) Complete this word equation to summarise the reaction. **(1 mark)**

> Sodium hydrogen → .... + ... + ...
> carbonate

b) Which everyday substance, usually found in the kitchen, contains sodium hydrogen carbonate? **(1 mark)**

....

....

....

....

2 When calcium carbonate is heated it reacts to form calcium oxide and carbon dioxide.

a) Complete the word equation to summarise this reaction. **(2 marks)**

> .... → .... + carbon dioxide

b) Explain why this reaction could be described as an example of thermal decomposition. **(2 marks)**

....

....

....

....

3 Limestone can be made into slaked lime.

The diagram below shows the steps involved in making slaked lime.

| limestone | Step 1 | quicklime CaO | Step 2 | slaked lime Ca(OH)$_2$ |

a) What is the formula of calcium carbonate? **(1 mark)**

....

....

b) In step 1 the calcium carbonate is heated to produce quicklime. Name the other product of this reaction. **(1 mark)**

....

....

c) What is the chemical name of quicklime? **(1 mark)**

....

....

d) In step 2 which substance is added to quicklime to produce slaked lime? **(1 mark)**

....

....

e) What is the chemical name of slaked lime? **(1 mark)**

....

....

How well did you do? ✗ 0-10 Try again   11-16 Getting there   17-21 Good work   22-26 Excellent! ✓

Fuels

A

1 **Which of these substances could be called fossil fuels?** (1 mark)

a) coal and uranium ☐
b) solar and wind ☐
c) wave and coal ☐
d) coal and oil ☐

2 **What is crude oil a mixture of?** (1 mark)

a) plastics ☐
b) elements ☐
c) hydrocarbons ☐
d) solids ☐

3 **Which of these descriptions best describes a short chain hydrocarbon?** (1 mark)

a) runny, hard to ignite and has a high boiling point ☐
b) viscous, hard to ignite and has a high boiling point ☐
c) viscous, easy to ignite and has a low boiling point ☐
d) runny, easy to ignite and has a low boiling point ☐

4 **What are groups of hydrocarbon molecules with a similar number of carbon atoms called?** (1 mark)

a) groups ☐
b) families ☐
c) sections ☐
d) fractions ☐

5 **What is formed by the cracking of hydrocarbon molecules?** (1 mark)

a) oxygen ☐
b) shorter hydrocarbon molecules ☐
c) longer hydrocarbon molecules ☐
d) pure carbon ☐

B

1 **True or false?** (5 marks)

	true	false
a) Crude oil is a mixture of hydrocarbons.	☐	☐
b) Hydrocarbons contain atoms of hydrogen, carbon and oxygen only.	☐	☐
c) A hydrocarbon molecule with 15 carbon atoms is found in the petrol fraction.	☐	☐
d) Fractional distillation can be used to separate a mixture of hydrocarbons.	☐	☐
e) Long chain hydrocarbon molecules are useful fuels.	☐	☐

2 **Cross out the incorrect word/phrase in the following sentences.** (5 marks)

a) Crude oil can be separated by filtering/fractional distillation.
b) Crude oil is a mixture/pure substance.
c) Long/short hydrocarbon molecules reach the top of the fractionating column before they condense.
d) Long hydrocarbon molecules are useful/not useful as fuels.
e) Long hydrocarbon molecules can be broken down into smaller hydrocarbon molecules by cracking/distilling.

C

1 Crude oil is extracted from the Earth's crust. It is a mixture of many substances; the most important ones are called hydrocarbons. Propane is obtained from crude oil. It can be used as a fuel. This diagram represents one molecule of propane.

$$H - \overset{\displaystyle\overset{H}{|}}{\underset{\displaystyle\underset{H}{|}}{C}} - \overset{\displaystyle\overset{H}{|}}{\underset{\displaystyle\underset{H}{|}}{C}} - \overset{\displaystyle\overset{H}{|}}{\underset{\displaystyle\underset{H}{|}}{C}} - H$$

a) Which of these options shows the formula of propane? Tick one box. **(1 mark)**

CH_4 ☐

C_2H_6 ☐

C_3H_6 ☐

C_3H_8 ☐

b) Is propane a hydrocarbon?
Explain your answer. **(1 mark)**

....

....

....

....

....

....

....

....

c) Crude oil can be separated using fractional distillation. The large hydrocarbon molecules separated during fractional distillation are not very useful. These large hydrocarbon molecules can be broken down into smaller, more useful hydrocarbons by cracking. Cracking is an example of a thermal decomposition reaction. Explain why this reaction can be described as an example of thermal decomposition. **(1 mark)**

....

....

....

....

....

....

d) A decane molecule can be split into two smaller molecules by cracking.

This reaction can be summarised by the equation

Decane → octane + ethene

i) How is octane used? **(1 mark)**

....

....

....

ii) How is ethene used? **(1 mark)**

....

....

....

Organic families

A

1 How many bonds do carbon atoms form? **(1 mark)**

a) 1
b) 2
c) 3
d) 4

2 What is the name for saturated hydrocarbons? **(1 mark)**

a) alkenes
b) alkanes
c) alcohols
d) esters

3 What is the test for an alkene? **(1 mark)**

a) It has carbon and hydrogen atoms only.
b) It can be burnt.
c) It decolourises bromine water.
d) It does not react with bromine water.

4 Which of these molecules is an alkane? **(1 mark)**

a) CH_4
b) C_2H_5OH
c) C_2H_4
d) CH_3OH

5 Which of these molecules is an alkene? **(1 mark)**

a) CH_4
b) C_2H_5OH
c) C_2H_4
d) CH_3OH

B

1 Name these hydrocarbons. **(4 marks)**

a)
H
|
H – C – H
|
H

b)
H H H
| | |
H – C – C – C – H
| | |
H H H

c)
H   H
  \   /
   C = C
  /   \
H   H

d)
H H H
| | |
H – C = C – C – H
              |
              H

....

....

2 a) Which of these hydrocarbons belong to the alkene family? **(5 marks)**

....

b) Which of these hydrocarbons belong to the alkane family?

....

c) Which of these hydrocarbons could be described as unsaturated?

....

d) Which of these hydrocarbons would react with bromine water?

....

e) Describe the colour change that takes place when bromine water reacts with an alkene.

....

C

1 Carbon atoms form four bonds with other atoms.

This means that carbon atoms can be made into an enormous range of compounds.

These three molecules all contain atoms of the element carbon.

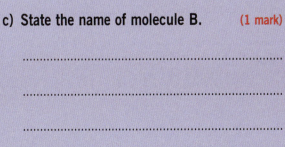

molecule A

molecule B

molecule C

a) Is molecule A a hydrocarbon? Explain your answer. **(1 mark)**

....

....

....

b) Is molecule B saturated? Explain your answer. **(1 mark)**

....

....

....

c) State the name of molecule B. **(1 mark)**

....

....

....

d) Which family of organic compounds does molecule C belong to? **(1 mark)**

....

....

....

e) State the name of molecule C. **(1 mark)**

....

....

....

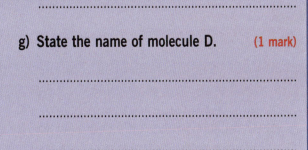

molecule D

f) Which family of organic compounds does molecule D belong to? **(1 mark)**

....

....

....

g) State the name of molecule D. **(1 mark)**

....

....

....

Vegetable oils

A

1 Which vitamins can we
get from plant oils? **(1 mark)**

a) C and A ☐
b) A and D ☐
c) C and B ☐
d) A and K ☐

2 Oils can be extracted from plants.
Which of these plant parts are least
likely to yield oils? **(1 mark)**

a) fruits ☐
b) stalks ☐
c) seeds ☐
d) nuts ☐

3 Which of these ways of cooking potatoes
would you expect to produce foods
with the highest energy content? **(1 mark)**

a) baking ☐
b) boiling ☐
c) frying ☐
d) microwaving ☐

4 Which of these foods
contains plant oil? **(1 mark)**

a) bacon ☐
b) olive oil ☐
c) cream ☐
d) butter ☐

5 What is the catalyst used in the
hydrogenation of plant oils? **(1 mark)**

a) nickel ☐
b) platinum ☐
c) gold ☐
d) sodium ☐

B

1 True or false? **(5 marks)**

	true	false
a) Salad dressing is an example of an emulsion.	☐	☐
b) Salad dressing is a mixture of alcohol and water.	☐	☐
c) Hydrophobic means attracted to water.	☐	☐
d) Hydrophilic means attracted to water.	☐	☐
e) Emulsifiers are molecules that help to keep the oil and water in mayonnaise mixed together.	☐	☐

2 Complete the table to show the name of the type of chemical added to food.
The first one is done for you. **(4 marks)**

Name of the chemical	Description of the chemical
colours	added to make the food look more attractive.
a)	added to decrease the amount of sugar used.
b)	added to enhance taste.
c)	added to help ingredients mix together.
d)	chemicals that are approved for use throughout the EU.

C

1 This label was taken from the side of a tub of low fat spread.

Ingredients:
Vegetable oils, hydrogenated vegetable oils, water, salt, emulsifier, preservative, colours, Vitamins A and D.

a) From which parts of a plant can vegetable oil be extracted? Tick two boxes. **(2 marks)**

Roots ☐
Seeds ☐
Leaves ☐
Nuts ☐
Flowers ☐

b) Why should people be careful about the amount of fat that they eat? **(1 mark)**

....

....

....

....

c) Animal fats are usually solid at room temperature. Vegetable fats are usually liquid at room temperature.

Vegetable oil is liquid

Golden sunflower oil

Butter is solid

BUTTER

i) What type of bond is present in vegetable fats that mean that they are liquid at room temperature? **(1 mark)**

....

....

....

....

....

ii) We can test for this bond using bromine water. If the bond is present what would you see when bromine water is added? **(1 mark)**

....

....

....

....

....

d) Emulsifiers are molecules that help water and oil to mix. One end of an emulsifier molecule is attracted to oil. What is the other end of the molecule attracted to? **(1 mark)**

....

....

....

....

....

How well did you do? ✗ **0-8** Try again   **9-12** Getting there   **13-16** Good work   **17-20** Excellent! ✓

Plastics

A

1 **What is the monomer used to make polythene?** (1 mark)

a) ethane ☐
b) ethene ☐
c) propene ☐
d) polymer ☐

2 **What is the monomer used to produce PVC?** (1 mark)

a) ethane ☐
b) chloroethene ☐
c) ethene ☐
d) tetrafluoroethene ☐

3 **What is the monomer used to produce Teflon?** (1 mark)

a) ethane ☐
b) chloroethene ☐
c) ethene ☐
d) tetrafluoroethene ☐

4 **What is the monomer used to produce polypropene?** (1 mark)

a) ethane ☐
b) ethene ☐
c) propene ☐
d) polymer ☐

5 **What does 'poly' mean?** (1 mark)

a) one ☐
b) two ☐
c) many ☐
d) ten ☐

B

1 **The diagrams opposite show three different hydrocarbons.**

molecule A

$$H-\overset{\displaystyle H}{\underset{\displaystyle H}{C}}-H$$

molecule B

$$H-\overset{\displaystyle H}{\underset{\displaystyle H}{C}}-\overset{\displaystyle H}{\underset{\displaystyle H}{C}}-\overset{\displaystyle H}{\underset{\displaystyle H}{C}}-H$$

molecule C

$$H-\overset{\displaystyle H}{C}=\overset{\displaystyle H}{C}-\overset{\displaystyle H}{\underset{\displaystyle H}{C}}-H$$

a) Which of these hydrocarbons represents an unsaturated hydrocarbon? (1 mark)

b) What is the name of molecule A? (1 mark)

c) Lot of molecules of 'molecule C' could be joined together to form a polymer. What is the name of this polymer? (1 mark)

2 **True or false?** (5 marks)

	true	false
a) Ethane is a monomer that can be made into polythene.	☐	☐
b) 'Poly' means three.	☐	☐
c) PVC is useful because it is flexible.	☐	☐
d) Ethene is made into polythene by heating many ethene molecules with a catalyst under high pressure.	☐	☐
e) The monomers used to make addition polymers have double bonds.	☐	☐

C

1 Polytetrafluoroethene, PTFE can be used to make non-stick saucepans.

PTFE is made by the polymerisation of tetrafluoroethene.

a) Is PTFE a hydrocarbon?
   Explain your answer. **(1 mark)**

....

....

....

....

b) PTFE is unsaturated.
   What does 'unsaturated' mean? **(1 mark)**

....

....

....

....

c) How could you prove that PTFE is unsaturated?

   i) What would you add? **(1 mark)**

....

....

....

   ii) What would you see? **(1 mark)**

....

....

....

....

d) Which of these equations shows the polymerisation of tetrafluoroethene? **(1 mark)**

i)
$$n \; \overset{\text{Cl}\;\;\text{H}}{\underset{\text{H}\;\;\text{H}}{C = C}} \xrightarrow[\text{cat}]{\text{temp}} \left(\overset{\text{Cl}\;\;\text{H}}{\underset{\text{H}\;\;\text{H}}{-C-C-}} \right)_n$$

ii)
$$n \; \overset{\text{F}\;\;\text{F}}{\underset{\text{F}\;\;\text{F}}{C = C}} \xrightarrow[\text{cat}]{\text{temp}} \left(\overset{\text{F}\;\;\text{F}}{\underset{\text{F}\;\;\text{F}}{-C-C-}} \right)_n$$

iii)
$$n \; \overset{\text{CH}_3\;\;\text{H}}{\underset{\text{H}\;\;\text{H}}{C = C}} \xrightarrow[\text{cat}]{\text{temp}} \left(\overset{\text{CH}_3\;\;\text{H}}{\underset{\text{H}\;\;\text{H}}{-C-C-}} \right)_n$$

e) When they are thrown away objects made from PTFE do not break down. Why don't they break down? **(1 mark)**

....

....

....

....

....

How well did you do? ✗ **0-8** Try again   **9-11** Getting there   **12-15** Good work   **16-19** Excellent! ✓

Ethanol

A

1 Which family of compounds does ethanol belong to? **(1 mark)**

a) alkanes ☐
b) alkenes ☐
c) carboxylic acids ☐
d) alcohols ☐

2 Alcohol can be made from glucose which of these raw materials does NOT contain glucose? **(1 mark)**

a) gooseberries ☐
b) apples ☐
c) grapes ☐
d) salt ☐

3 What is the name of the process in which glucose is converted to alcohol? **(1 mark)**

a) distillation ☐
b) filtration ☐
c) fermentation ☐
d) cracking ☐

4 What is the catalyst used when glucose is converted to ethanol? **(1 mark)**

a) sugar ☐
b) heat ☐
c) yeast ☐
d) bacteria ☐

5 What are the conditions used in to make industrial alcohol? **(1 mark)**

a) yeast and room temperature ☐
b) yeast and phosphoric acid ☐
c) yeast and a temperature of 300°C ☐
d) phosphoric acid and a temperature of 300°C ☐

B

1 Complete the following passage. **(12 marks)**

yeast	less	fermentation	renewable	low	properties
alcohol	dissolve	evaporates	cereals	dioxide	fuel

Ethanol is a member of the .... family. Ethanol has many useful .... Many perfumes contain ethanol. It is a good solvent so things .... well in ethanol. Ethanol also has quite a .... boiling point so it quickly .... from the skin.

In some countries ethanol made from sugar beet or sugarcane is used as a .... Ethanol is a .... energy source. Unfortunately, burning alcohol releases .... energy than burning petrol.

Traditionally, alcohol has been produced by .... During fermentation glucose found in fruit, vegetables and .... is converted into alcohol and carbon .... The reaction is catalysed by ....

2 True or false? **(4 marks)**

	true	false
a) Methanol is toxic.	☐	☐
b) During fermentation ethanol is produced faster at 25°C than at 5°C.	☐	☐
c) Yeast contains an enzyme.	☐	☐
d) Ethane is used in the industrial production of ethanol.	☐	☐

C

1 A chemical reaction can convert glucose into two useful new substances.

a) Name these two substances. **(2 marks)**

....

....

....

....

2 This diagram shows the apparatus used to produce alcohol in the laboratory.

a) What is this process called? **(1 mark)**

....

....

....

....

b) What is the name of the substance added to catalyse the reaction? **(1 mark)**

....

....

....

....

c) Alcohol can also be produced industrially. Long chain hydrocarbons are not useful as fuels. The cracking of long hydrocarbons produces shorter hydrocarbons that are useful as fuels and ethene molecules. Ethene molecules can be reacted with another chemical to produce alcohol. Complete the word equation by adding the name of the other reactant in this reaction. **(1 mark)**

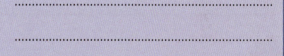

ethene + ... → ethanol

3 This is a diagram showing the structure of ethanol.

$$H-\overset{\displaystyle\overset{H}{|}}{\underset{\displaystyle\underset{H}{|}}{C}}-\overset{\displaystyle\overset{H}{|}}{\underset{\displaystyle\underset{H}{|}}{C}}-O-H$$

a) Which organic family does ethanol belong to? **(1 mark)**

....

....

b) Why is ethanol used in aftershaves? **(1 mark)**

....

....

....

Evolution of the atmosphere

A

1 **Which gases comprised the Earth's early atmosphere?** *(1 mark)*

a) carbon dioxide, steam, ammonia and methane ❑

b) oxygen, nitrogen and trace amounts of carbon dioxide, water vapour and noble gases ❑

c) carbon monoxide and water vapour ❑

d) oxygen, nitrogen and carbon monoxide ❑

2 **For how long have the levels of carbon dioxide in the atmosphere been rising?** *(1 mark)*

a) since the last general election ❑

b) the last two hundred years (since the industrial revolution) ❑

c) since the middle ages ❑

d) over the last twenty years ❑

3 **How is the extra carbon dioxide in the atmosphere produced?** *(1 mark)*

a) when fossil fuels are formed ❑

b) when fossil fuels are burnt ❑

c) when fossil fuels are mined ❑

d) when fossil fuels are buried ❑

4 **Which gases comprise the Earth's atmosphere today?** *(1 mark)*

a) carbon dioxide, steam, ammonia and methane ❑

b) oxygen, nitrogen and trace amounts of carbon dioxide, water vapour and noble gases ❑

c) carbon monoxide and water vapour ❑

d) oxygen, nitrogen and carbon monoxide ❑

5 **Why is the ozone layer useful to us?** *(1 mark)*

a) it stops acid rain ❑

b) it prevents global warming ❑

c) it filters out harmful UV rays ❑

d) it helps to reduce car crime ❑

B

1 **Place these events in order to show how the Earth's atmosphere has evolved.** *(5 marks)*

a) Plants evolved and soon covered most of the Earth. ❑

b) Carbon dioxide became locked up as carbonate minerals and in fossil fuels. ❑

c) Enormous volcanic activity produced carbon dioxide, steam, ammonia and methane. ❑

d) Plants removed carbon dioxide and produced oxygen. ❑

e) The water vapour (steam) condensed to form the early oceans. ❑

2 **True or false?** *(5 marks)*

	true	false
a) Nitrogen is produced by denitrifying bacteria.	❑	❑
b) Earth's early atmosphere was similar to the modern day atmosphere of Neptune and Pluto.	❑	❑
c) Ammonia is a gas.	❑	❑
d) The evolution of green plants lead to a decrease in the amount of oxygen and an increase in the amount of carbon dioxide in the atmosphere.	❑	❑
e) Carbon dioxide can be removed from the atmosphere by the reaction between carbon dioxide and seawater .	❑	❑

C

1 This question is about how the amount of oxygen in the atmosphere has increased over time. Use the words and phrases in the box below to complete the sentences. **(4 marks)**

carbon dioxide	ozone
harmful	plants

As a) .... evolved the amount of oxygen

in the atmosphere increased. These plants

grew well in the b) .... rich atmosphere.

As the amount of oxygen increased an c)

.... layer developed. This layer filtered

out d) .... ultraviolet rays.

2 Over time the composition of the Earth's atmosphere has changed. Scientists believe that the Earth is 4.5 billion years old. During the first billion years of the Earth's history there was enormous volcanic activity. These volcanoes released large amounts of gas which formed the Earth's early atmosphere.

a) Which of these gases was NOT produced in large quantities by these volcanoes? Tick one box. **(1 mark)**

methane ☐

CFCs ☐

ammonia ☐

carbon dioxide ☐

steam ☐

b) The Earth's early atmosphere mainly consisted of carbon dioxide. Name a planet in the Solar system which has an atmosphere today similar to the Earth's early atmosphere. **(1 mark)**

....

....

....

....

c) What did the steam released by the volcanoes eventually produce? **(1 mark)**

....

....

....

....

d) Why did the amount of oxygen in the atmosphere eventually increase? **(1 mark)**

....

....

....

....

e) The amount of nitrogen in the Earth's atmosphere has also increased over time. Some nitrogen was produced when ammonia reacted with oxygen. How else was nitrogen produced? **(1 mark)**

....

....

....

....

Pollution of the atmosphere

A

1 Which of these elements may be found in small amounts in fossil fuels? **(1 mark)**

- a) tin ☐
- b) sulphur ☐
- c) gold ☐
- d) silicon ☐

2 Which of these descriptions best describes the gas carbon monoxide? **(1 mark)**

- a) green and dense ☐
- b) colourless and smelly ☐
- c) colourless, odourless and very poisonous ☐
- d) poisonous and violet ☐

3 When fuels are burnt why is carbon monoxide sometimes produced? **(1 mark)**

- a) insufficient supply of nitrogen ☐
- b) insufficient supply of fuel ☐
- c) insufficient supply of heat ☐
- d) insufficient supply of oxygen ☐

4 Which of these gases is linked with 'acid rain' **(1 mark)**

- a) carbon monoxide ☐
- b) carbon dioxide ☐
- c) sulphur dioxide ☐
- d) nitrogen ☐

5 Which of these events could be a consequence of global warming? **(1 mark)**

- a) changes to the ozone layer ☐
- b) icecaps could melt and cause massive flooding ☐
- c) acid rain ☐
- d) faulty gas appliances ☐

B

1 Complete the following passage. **(7 marks)**

serviced   monoxide   hydrogen   poisonous   oxygen   vapour   kills

Hydrocarbon fuels contain .... and carbon. When these fuels are burnt the gases carbon dioxide and water .... are produced and energy is released. Sometimes fuels are burnt in a poor supply of .... When this happens the gas carbon .... may also be produced. Carbon monoxide is very dangerous. It is colourless, odourless and very .... This gas .... many people every year. Faulty gas appliances are particularly dangerous so it is important that they are regularly ....

2 True or false? **(5 marks)**

	true	false
a) Global dimming is caused by smoke particles.	☐	☐
b) Global warming is caused by seawater rises.	☐	☐
c) Incomplete combustion of fuels produces soot.	☐	☐
d) All scientists now accept that human activity is causing the Greenhouse effect.	☐	☐
e) Acid rain is caused by sulphur dioxide pollution.	☐	☐

C

1 Many fuels contain carbon. Complete the word equation to show what happens when carbon burns to form carbon dioxide.

a) Carbon + ... → carbon dioxide

(1 mark)

b) Humans are affecting the proportion of gases in the atmosphere.

i) How is the amount of carbon dioxide in the atmosphere changing? (1 mark)

....

....

....

ii) Why is the amount of carbon dioxide in the atmosphere changing? (1 mark)

....

....

....

....

c) Some fuels contain traces of sulphur. Complete the equation to show what happens when sulphur burns to form sulphur dioxide.

Sulphur + oxygen → .... (1 mark)

d) Which of these environmental problems could be caused by acid rain? Tick two boxes. (2 marks)

damage to statues ☐
global dimming ☐
damage to trees ☐
skin cancers ☐
changes to weather patterns ☐
increased sea levels ☐

e) Which of these environmental problems could be caused by increased levels of smoke particles in the atmosphere? Tick one box. (1 mark)

damage to statues ☐
global dimming ☐
damage to trees ☐
skin cancers ☐

2 a) Name the two elements found in hydrocarbon fuels. (2 marks)

....

....

....

....

....

....

b) When hydrocarbon fuels are burnt the gas carbon dioxide can be made. Which of these environmental problems could be caused by increased levels of carbon dioxide in the atmosphere? Tick one box. (1 mark)

damage to trees ☐
increased sea levels ☐
global dimming ☐
skin cancers ☐

How well did you do? ✗ 0-10 Try again   11-14 Getting there   15-21 Good work   22-27 Excellent! ✓

Pollution of the environment

A

1 Which of these gases can be produced when PVC is burnt? (1 mark)

a) hydrogen chloride ☐
b) chlorine ☐
c) cyanide ☐
d) bromine ☐

2 Which of these substances could be described as non-biodegradable? (1 mark)

a) bananas ☐
b) leaves ☐
c) newspapers ☐
d) most plastics ☐

3 Which of these issues could be an advantage of limestone quarrying in an area? (1 mark)

a) destruction of animal habitats ☐
b) new jobs ☐
c) many heavy lorries ☐
d) scarring of the landscape ☐

4 Which of these statements is true of non-biodegradable plastics? (1 mark)

a) they are reactive ☐
b) they do not rot away ☐
c) they react with water ☐
d) they react with oxygen in the air ☐

5 What is the main ore of aluminium? (1 mark)

a) aluminium sulphide ☐
b) haematite ☐
c) bauxite ☐
d) cryolite ☐

B

1 Complete this passage by crossing out the incorrect word or phrase. (5 marks)

a) Aluminium ore is extracted from large/small open cast mines.
b) Trees have to be cut down near aluminium ore mines to build airports/roads.
c) By recycling aluminium objects in this country we can increase/decrease the speed at which our tips are filled up.
d) The area around mines is improved/polluted by litter and oil.
e) Aluminium/sodium is used to make drinks cans.

2 Complete the following passage. (7 marks)

Plastics are very stable so they do not .... easily. This makes plastics very useful. Most plastics do not react with water or with .... in the air. They are also .... which means that they are not decomposed by microbes. Some plastics can be .... however this can produce poisonous gases like ....

Scientists have recently developed a new range of .... plastics. These plastics are much easy to dispose of because they will eventually .... away.

C

1 This question is about the extraction of limestone rock. Use the words in the box below to complete the sentences. **(4 marks)**

landscape	quarries
pollution	calcium carbonate

Limestone is a type of sedimentary rock. The

main chemical compound in limestone is a)

.... Limestone is extracted from b)

.... in large quantities. This can scar the

c) .... and cause noise d) ....

2 This question is about plastics. Use the words in the box below to complete the sentences.

(4 marks)

hydrogen	chloride	microbes
oxygen	unreactive	

Most plastics are very a) .... They do

not react with b) .... in the air or with

living .... Some plastics can be

burnt. Burning the plastic PVC produces

d) ....

3 Aluminium can be extracted from the mineral bauxite. Bauxite is often found in environmentally sensitive areas like the Amazonian rain forest.

a) Suggest one way in which a new bauxite quarry could be of benefit to people who live near to the mine. **(1 mark)**

....

....

....

....

b) Suggest one way in which a new bauxite quarry could damage the local environment. **(1 mark)**

....

....

....

....

Evidence for plate tectonics

A

1 Which of these layers is found at the centre of the Earth? **(1 mark)**

a) outer core
b) inner core
c) mantle
d) crust

2 What is included in the Earth's lithosphere? **(1 mark)**

a) mantle and outer core
b) inner and outer core
c) crust and upper mantle
d) crust and outer core

3 What do we think causes the convection currents that drives the movement of the Earth's plates? **(1 mark)**

a) earthquakes
b) icelandic power stations
c) natural radioactive decay
d) magnetism

4 What do scientists think that the Earth's core is made of? **(1 mark)**

a) carbon and silicon
b) iron and nickel
c) iron and silicon
d) silicon and oxygen

5 Which elements are found in the highest amounts in the Earth's crust? **(1 mark)**

a) silicon, iron and magnesium
b) silicon, oxygen and iron
c) magnesium and nickel
d) silicon, aluminium and oxygen

B

1 Label the diagram to show the layered structure of the Earth. **(5 marks)**

a)
b)
c)
d)
e)

2 True or false? **(5 marks)**

	true	false
a) The Earth's lithosphere is split into three plates.	☐	☐
b) Oceanic crust is mainly basalt.	☐	☐
c) The outer and inner cores are liquid.	☐	☐
d) The density of rocks increases with depth.	☐	☐
e) The Earth's mantle slowly flows.	☐	☐

Letts
and
LONSDALE

GCSE
Success

Workbook
Answer
Booklet

Science
Foundation

Brian Arnold • Elaine Gill • Emma Poole

Answers

Biology

Pages 4–5 A balanced diet and nutrition

A
1. b   2. b   3. b   4. a   5. b

B

1.

Nutrient	Found in	Used for
carbohydrate	cereals	energy
fibre	plants	moving food in gut
water	all food and drink	cools us down
protein	lean meat	growth/repair of cells

2. Fats – energy store, make membranes, warmth
Vitamins – all reactions

C
1. a) 400 kJ
   b) Tom's tomatoes – contains less energy/carbohydrates/fats
   c) more protein
2. a) 1000 kJ
   b) growing
   c) needs energy for growing embryo/foetus/baby
   d) anorexia
   e) reduced resistance to infection/pale, papery skin/in women, irregular periods.

Pages 6–7 The nervous system

A
1. b   2. c   3. c   4. d   5. c

B

1.

Stimulus	Sense	Sense organ
light	sight	eye
chemicals	taste	taste buds
sound waves	hearing	ears
pressure/ temperature	touch	skin
chemicals	smell	nose

2. voluntary; conscious; learned; talking; involuntary; reflex.

C
1. a) A – cell membrane
      B – nucleus
      C – nucleus
   b) longer/bigger

2.

| stimulus |
| receptor |
| sensory neurone |
| relay neurone |
| motor neurone |
| effector |
| response |

Pages 8–9 The eye

A
1. c   2. b   3. b   4. b   5. b

B

1.

Cornea	helps focus the image
Lens	a hole that allows light through (in front of the lens)
Muscular iris	the protective, white outer layer of the eye
Optic nerve	contains light sensitive cells
Pupil	controls how much light enters the eye
Retina	transparent window in the front of the eye
Sclera	receives nerve impulses from the retina and sends them to the brain

2. a) false
   b) true
   c) true

C
1. a) A – ciliary muscle
      B – suspensory ligament
      C – lens
   b) Short sight; eyeball too long/rays focus too soon in front of retina.
2. a)

B	D	E	A	C

3. circular, radial, smaller
4. A cornea
   B pupil
   C iris
   D sclera
   E optic nerve
   F retina
5. a) eyes on front of head – can judge distances and depth
   b) eyes on side of head – all round vision

Pages 10–11 The brain

A
1. c   2. d   3. a   4. d   5. d

B
1. a) stare into space, freeze, no convulsions
   b) convulsions/twitches/loss of consciousness
   c)

Disorder	Possible causes/ increases risk	Symptoms/facts
strokes	blood supply to brain or part of is stopped/high blood pressure/ smoking	paralysis/loss of speech/numbness/ vision/headache/ loss of dizziness
Parkinsons	unknown	tremors/rigidity/ difficulty walking/ poor balance
tumours	exposure to radiation/chemicals	uncontrollable growth of cells/puts pressure on brain/malignant, cancerous or benign

C
1. a)

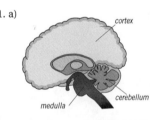

cortex

medulla

cerebellum

b)

Function	Part of brain
controlling balance	cerebellum
making decisions	cortex
controlling heart and breathing rates	medulla

c)

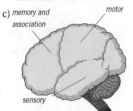

memory and association

motor

sensory

d) outer layer of brain divided down middle in two cerebral hemispheres

different areas have different functions
2. The brain works by sending **electrical impulses** received from the sense organs to the **muscles**. In mammals, the brain is complex and involves billions of **neurones** that allow learning by experience and behaviour.
The interaction between mammals and their **environment** results in nerve pathways forming in the brain. When mammals learn from experience, **pathways** in the brain become more likely to transmit impulses than others, which is why it is easier to **learn** through repetition.

Pages 12–13 Causes of disease

A
1. c   2. d   3. a   4. c   5. c

B

1.

Feature	Bacteria	Viruses
cell wall	✓	✗
protein coat	✗	✓
respire, feed and move	✓	✗
reproduce inside living cells	✗	✓

2. chlorophyll (chloroplast) and vacuole.
3. temperature/headache/loss of appetite/sickness.

C
1. a) sufferers cough or sneeze others breathe in bacteria from air
   b) antibiotics
      better public health
      vaccination
2.

Disease	Cause
anaemia	lack of iron
cancer	cell division out of control
red-green colour blindness	inherited
scurvy	lack of vitamin C

3. **Contact** with infected people, animals or objects used by infected people, e.g. athlete's foot, chicken pox and measles.
Through the **air**, e.g. flu, colds and pneumonia.
Through infected **food and drink**, e.g. cholera from infected drinking water and salmonella food poisoning.

Pages 14–15 Defence against disease

A
1. d   2. c   3. b   4. d   5. a

B
1. a) viruses
   b) new strains of bacteria, bacteria resistant to current antibiotics
2. white blood cells
   antitoxins
   phagocytes
   lymphocytes

C
1. a) barrier
      produces sebum
      which is an antiseptic, waterproof/liquid
   b) white blood cells/phagocytes engulf/eat microbes
2. a) antigens on mumps, microbe recognised as foreign, lymphocytes produce antibodies, clump microbes together, phagocytes engulf them
   b) dead/harmless/weakened microbes are injected/by vaccination
   c) antibodies are injected

Pages 16–17 Drugs

A
1. a   2. b   3. b   4. b   5. c

B
1. Drugs are powerful **chemicals**; they alter the way the body works, often without you realising it. There are **useful** drugs such as penicillin and antibiotics, but these can be dangerous if misused. Drugs affect the **brain** and **nervous system**, which in turn affects **behaviour** and risk of infection.
2. a false
   b true
   c true

C
1. a) lowers it
   b) prevents mother's blood from carrying oxygen which deprives foetus of oxygen
2.

Drug	Organs
alcohol	1 brain
	2 liver
solvents	1 liver
	2 kidneys
	3 heart
painkillers	1 brain

3. bronchitis, emphysema, heart disease
4. a) nicotine
   b) carbon monoxide – prevents red blood cells from carrying oxygen
      tar – carcinogen

Pages 18–19 Homeostasis and diabetes

A
1. b   2. a   3. c   4. c   5. d

B
1. the kidney
2. blood glucose
   temperature
   salt/water content
3. i) A
   ii) C
   iii) B
   iv) D

C
1. glucose, hormones, high, insulin, liver, glycogen, normal, low, glucagon
2. a) protein
   b) urea
   c) reabsorbed in the kidney

Pages 20–21 The menstrual cycle

A
1. a   2. b   3. d   4. d   5. b

B

1.

Change	Boys	Girls
breasts develop		✓
genitals develop	✓	✓
hair grows under the arms	✓	✓
hair grows on the face and body	✓	
menstruation begins		✓
pubic hair grows	✓	✓
sperm production begins	✓	
voice deepens	✓	

2. prepare the uterus to receive a fertilised egg, breaks down uterus wall when fertilisation does not occur

C

1. a) i) C
   ii) B
   iii) A
   b) lining of uterus breaks down caused by lack of progesterone
   c) causes uterus lining to build up, stimulates egg development, stimulates ovulation
2. a) stimulate eggs to mature when female's own production is low
   b) can result in multiple egg release/births
3. maintains the uterus lining, menstruation
4. taking eggs and sperm from a female and male, fertilisation outside body, implant embryo into female uterus

Pages 22–23   Genetics and variation

A

1. c   2. c   3. d   4. d   5. a

B

1. a) any two inherited features e.g. gender, eye colour
   b) any two environmental factors e.g. scars, weight
2. a) true
   b) true
   c) true

C

1. a) temperature, moisture, type of soil, light nutrients,
   b) all the same colour, asexual reproduction
   c) clones
2. taking cuttings; grow the cuttings in different conditions; if colours all the same – not due to environment/if different – due to environment; repeat
3. eye colour, natural hair colour, blood group, inherited disease
4. sunlight, soil type, water, temperature

Pages 24–25   Genetics

A

1. b   2. c   3. c   4. b   5. a

B

1.

Definition	Word
different alleles	heterozygous
the stronger allele	dominant
the type of alleles the organism has	genotype
the weaker allele	recessive
what the organism looks like	phenotype
both alleles the same	homozygous

2. chromosomes, protein, disease, insulin

C

1. a)

	r	r
R	Rr/rR	Rr/rR
r	rr	rr

   b) i) round
      ii) round
      iii) round
      iv) wrinkled
2. a) using genetic engineering to treat a genetic disease
   b) insert the correct gene into body cells
   c) Many different body cells have the wrong gene so getting the cells to take up the gene would be hard. Cells would not multiply once the gene inserted so some still have the faulty gene.
3. a) medicine, agriculture
   b) We have developed plants that are resistant to pests and diseases.
   Plants that can grow in adverse environmental conditions.
   Tomatoes and other sorts of fruit are now able to stay fresher for longer.
   Animals are engineered to produce chemicals in their milk, such as drugs and human antibodies.

Pages 26–27   Inherited diseases

A

1. c   2. b   3. c   4. c   5. c

B

1. a) HH or Hh
   b) hh
   c) Hh
   d) hh
2. onset/symptoms do not show until 30–40 years old; by then person may already have children

C

1. a) FF/Ff
   b) Ff
   c) person who has the allele/gene does not suffer heterozygous
   d) 1 in 4/25%
   e) 2 in 4/1 in 2/50%
2. a) produce large amounts of thick, sticky mucus; mucus blocks air passages and digestive tubes; difficulty breathing; difficulty in absorbing food; chest infections
   b) physiotherapy, antibiotics
   c) Explain to them that there is a one in four chance of having a child with cystic fibrosis. They then have to decide if the risk of having children is too great.
3. a) adult bone marrow, human embryo, umbilical cord
   b) stem cells have the ability to divide and specialise into any tissue/cell
   c) funding, support, regulation

Pages 28–29   Selective breeding

A

1. c   2. a   3. b   4. a   5. a

B

1. a) They cannot cope with a change in environment and will all die. No alleles left to breed new varieties.
   b) to maintain species variation
2. a)

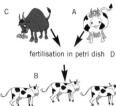

fertilisation in petri dish

   b) Sperms and eggs can be frozen and used later. Many offspring can be produced by one bull or one cow.

C

1.

S	P	T	Q	R

2. a) asexual
   b) All offspring will be identical.
3. a) People like eating large tasty strawberries.
   b) stage 1 – few cells taken from selected plant
   stage 2 – growth medium prepared
   stage 3 – cells placed in growth medium
   c) plants grown all year, plants grown quickly, plants grown cheaply, plants can be transported easily when small

Pages 30–31   Pyramids

A

1. b   2. b   3. d   4. b   5. d

B

1. a) grass
   b) Pyramid of numbers: largest box at bottom, smallest box at top. Labelled: grass at bottom, then rabbit, then owl on top.
   c) inverted pyramid smallest at bottom, biggest in middle labelled, rose at bottom, ladybird at top
   d) does not tell us the size of the organisms

C

1. a) pyramid = 1
   Labels correct = 3

   Hawk
   Voles
   Caterpillars
   Tree

   b) number of each organism multiplied by mass on one organism
2. a) 75 + 15 = 90% lost
   100 – 90 = 10. Answer = 10%
   b) respiration/heat
   c) urine/faeces
   d) food

Pages 32–33   Evolution

A

1. a   2. b   3. b   4. c   5. d

B

1. a) competition for food, predators, disease
2. a) Organisms produce more offspring than could possibly survive. Population numbers remain fairly constant despite this. All organisms in a species show variation. Some of these variations are inherited.
   b) There was a struggle for existence and the strongest/fittest survived to reproduce.

C

1. a) mutation
   b) easily seen, so eaten
   c) camouflaged
   d) increase in light colour decrease in dark colour
2. Animals die.
   The hard parts of animals that do not decay, form into a rock. Minerals gradually replace the softer parts of animals in areas where there is no oxygen, moisture or warmth.

Pages 34–35   Adaptation and competition

A

1. c   2. a   3. b   4. a   5. d

B

1.

Definition	Word
where organism lives	habitat
all one type of animal or plant	population
living things in the habitat	community
all the living things and their physical environment	ecosystem

2. The amount of food and water available.
   Predators or grazing - who may eat the animal or plant.
   Disease.
   Climate, temperature, floods, droughts and storms.
   Competition for space, mates, light, food and water.
   Human activity, such as pollution or destruction of habitats.
   Organisms will only live and reproduce where conditions are suitable.
3. a) An animal who hunts and kills another animal.
   b) A prey is the hunted animal.

C

1. It has a thick coat to keep in body heat as well as a layer of blubber for insulation.
   Its coat is white so that it can blend into its surroundings.
   Its fur is greasy so doesn't hold water after swimming.
   A polar bear has big feet to spread its weight on snow and ice.
   It also has big sharp claws to catch fish.
   It is good swimmer and runner to catch prey.
2. a) Population goes up and down. Increase and decrease slightly out of phase to the predator.

b) Population goes up and down. Increase and decrease slightly out of phase to the prey.
c) i when predator population is increasing
ii more are being eaten
d) i fewer predators
ii less being eaten

Pages 36–37 Environmental damage 1
A
1. d 2. c 3. b 4. a 5. b
B
1. intensive,
food,
pesticides and fertilisers,
pesticides
2. Trees are burned.
This produces CO_2 contributing to the greenhouse effect.
Without trees soil is easily washed away.
Over time deserts may form.
Trees transpire water which falls as rain in other areas.
The loss of large forests means some areas will face drought.
C
1. a) washed into rivers/lakes, water gets taken into plants, then plants eaten by animals
b) 5 ppm
c) DDT cannot be broken down, so is passed along food chain,
gets concentrated in grebe, levels toxic
2. a) predatory mites eat the red spider mites
b) biological
c) only kills red spider mites/pest, does not harm other species, no chemicals
d) predatory mite may become a pest;
takes time; does not remove all pests

Pages 38–39 Environmental damage 2
A
1. b
2. c
3. b
4. b
5. c
B
1. a) true
b) false
c) true
d) true
2. a) chlorofluorocarbons
b) aerosols, fridges, plastic foam
c) ozone layer develops holes
d) harmful UV rays reaching earth, increased risk of skin cancer
C
1. a) carbon dioxide, sulphur dioxide, nitrogen oxides
b) gases dissolve in water in clouds forming acids
c) i dissolves the stone
ii makes the water acidic, kills fish/living organisms
iii kills them
2. a) greenhouse effect
b) global warming

c) i) small
ii) more
iii) UV
iv) most
v) UV
vi) less

Pages 40–41 Ecology and classification
A
1. b 2. c 3. d 4. c 5. b
B
1. a) Kingdom
Phylum
Class
Order
Family
Genus
Species
b) group of organisms that can breed together to produce fertile offspring
2. fish, amphibians, reptiles, birds, mammals
C
1. a) P – beetle
Q – snail
R – slug
S – spider
2. a) A – quadrat
B – pooter
C – pitfall trap
b) add up numbers in quadrats = 60
average number per quadrat = 6
quadrat = square metre
field 10000 square metres
number of daisies = 10000 $\times$ 6 = 60000

Chemistry
Pages 42–43 Limestone
A
1. a 2. c 3. d 4. c 5. b
B
1. a) quicklime/calcium oxide/carbon dioxide
b) cement
c) mortar
d) slaked lime/calcium hydroxide/limewater
e) glass
2. a) true
b) false
c) false
d) true
e) false
C
1. a) sodium carbonate + carbon dioxide + water
b) baking powder
2. a) calcium carbonate $\rightarrow$ calcium oxide
b) We are using heat to break down the calcium carbonate into simpler substances.
3. a) $CaCO_3$
b) carbon dioxide
c) calcium oxide
d) water
e) calcium hydroxide

Pages 44–45 Fuels
A
1. d 2. c 3. d 4. d 5. b
B
1. a) true
b) false
c) false
d) true
e) false
2. a) fractional distillation
b) mixture
c) short
d) not useful
e) cracking

C
1. a) C_3H_8
b) Yes, it contains carbon and hydrogen only.
c) Heat is being used to breakdown large molecules into simpler substances.
d) i) As a fuel/in petrol
ii) to make plastics e.g. polyethene

Pages 46–47 Organic families
A
1. d 2. b 3. c 4. a 5. c
B
1. a) methane
b) propane
c) ethene
d) propene
2. a) c and d
b) a and b
c) c and d
d) c and d
e) Bromine water decolourises or turns from orange/brown to colourless.
C
1. a) No, it contains hydrogen, carbon and oxygen.
b) Yes, it has no double bonds.
c) propane
d) alkene
e) propene
f) alkene
g) butene

Pages 48–49 Vegetable oils
A
1. b 2. b 3. c 4. b 5. a
B
1. a) true
b) false
c) false
d) true
e) true
2. a) sweeteners
b) flavours
c) emulsifiers
d) E-numbers
C
1. a) seeds, nuts
b) get fat/heart disease/raised cholesterol levels
c) i) C=C
ii) decolourises
d) water

Pages 50–51 Plastics
A
1. b 2. b 3. d 4. c 5. c
B
1. a) molecule c
b) methane
c) polypropene
2. a) false
b) false
c) false
d) true
e) true
C
1. a) No, it does not contain hydrogen and it does contain fluorine.
b) It has double bonds.
c) i) add bromine water
ii) decolourises/ orange or brown to colourless
d) ii)
e) non-biodegradable/ unreactive/does not react with water or oxygen

Pages 52–53 Ethanol
A
1. d 2. d 3. c 4. c 5. d

B
1. alcohol, properties, dissolve, low, evaporates, fuel, renewable, less, fermentation, cereals, dioxide, yeast
2. a) true
b) true
c) true
d) false
C
1. a) ethanol and carbon dioxide
2. a) fermentation
b) yeast
c) water
3. a) Alcohols
b) It is a good solvent/it has a low boiling point

Pages 54–55 Evolution of the atmosphere
A
1. a 2. b 3. b 4. b 5. c
B
1. c, e, a, d, b
2. a) true
b) false
c) true
d) false
e) true
C
1. a) plants
b) carbon dioxide
c) ozone
d) harmful
2. a) CFCs
b) Venus or Mars
c) (early) oceans
d) plants evolved
e) living organisms like denitrifying bacteria

Pages 56–57 Pollution of the atmosphere
A
1. b 2. c 3. d 4. c 5. b
B
1. hydrogen, vapour, oxygen, monoxide, poisonous, kills, serviced
2. a) true
b) false
c) true
d) false
e) true
C
1. a) oxygen
b) i) it is increasing
ii) burning more fossil fuels
c) sulphur dioxide
d) damage to statues damage to trees
e) global dimming
2. a) hydrogen and carbon
b) increased sea levels

Pages 58–59 Pollution of the environment
A
1. a 2. d 3. b 4. b 5. c
B
1. a) large
b) roads
c) decrease
d) polluted
e) aluminium
2. react
oxygen
non-biodegradable
burnt
hydrogen chloride
biodegradable
rot
C
1. a) calcium carbonate
b) quarries
c) landscape
d) pollution
2. a) unreactive
b) oxygen

c) microbes
d) hydrogen chloride
3. a) jobs/money/better
roads/better facilities
b) trees have to be cut down
for the quarry or access
roads/loss of land/oil or
litter or noise
pollution/destroy local
plant or animal life

Pages 60–61   Evidence for plate tectonics
A
1. b   2. c   3. c   4. b   5. d
B
1. a) crust
b) mantle
c) lithosphere
d) outer core
e) inner core

2. a) false
b) true
c) false
d) true
e) true
C
1. a) crust and upper mantle
b) jigsaw fit of coasts, similar
fossil records, similar rock
strata
c) convection currents
caused by natural
radioactive decay
d) a few cms per year
e) inner core - solid, outer
core-liquid
f) crust

Pages 62–63   Consequences of plate tectonics
A
1. d   2. b   3. a   4. b   5. c
B
1. a) oceanic plate
b) continental plate
c) volcano
d) fold mountains
e) melting

2. boundaries, slide, San
Andreas, plates, stuck, forces,
earthquake
C
1. a) i)   mantle
ii) outer core
iii) inner core
iv) crust
b) lithosphere
c) metamorphic
d) it is denser
e) West coast of South
America
f) earthquake/volcano/
tsunami
g) there are too many factors
involved

Pages 64–65   Extraction of iron
A
1. c   2. b   3. b   4. d   5. d
B
1. haematite
oxide
silicon dioxide
limestone
furnace
slag
slag
density
2. a) false
b) true
c) false
d) true
e) false
C
1. a) limestone and coke
b) haematite
c) hot air/oxygen

d) carbon dioxide
e) reduction
f) reacts with silica
impurities to form slag
2. carbon

Pages 66–67   Iron and steel
A
1. b   2. c   3. b   4. d   5. b
B
1. a) oxygen and water
b) more
c) non-metallic
d) harder
e) different sizes
2. carbon, cast, rust, brittle,
drain covers etc, carbon,
softer, shape, atoms,
pass/slip, gates, etc.
C
1. a) water and oxygen
b) stops water and oxygen
reaching the iron
2. a) atoms are the same size
and have a regular
arrangement (1 mark) and
are labelled as iron atoms
(1 mark)
b) Cast iron: in cast iron the
atoms are different sizes so
they do not have a regular
arrangement and the
layers cannot pass easily
over each other.
c) alloy
d) strong
e) the more carbon the
harder they are to shape

Pages 68–69   Aluminium
A
1. d   2. c   3. d   4. b   5. a
B
1. a) negative ion
b) positive electrode
c) negative electrode
d) positive ion
2. a) false
b) true
c) false
d) true
e) true
C
1. a) electrolysis
b) bauxite
c) cryolite
d) It has a lower melting
point and bauxite dissolves
in molten cryolite.
e) reduction
f) positive
g) blue section labelled
h) graphite/carbon
i) The oxygen that is formed
there reacts with the
carbon to form carbon
dioxide/carbon monoxide,
so the graphite electrode is
eaten away.

Pages 70–71   Titanium
A
1. a   2. b   3. a   4. d   5. d
B
1. a) true
b) false
c) true
d) true
e) false
2. a) soft
b) nickel and titanium
c) less reactive
d) return to their original shape
e) more reactive
C
1. a) rutile
b) alloy
c) magnesium
d) nitinol
2. a) high density

b)
magnesium + titanium → magnesium + titanium
chloride   chloride
c) magnesium displaces
titanium from titanium
chloride
d) to stop the titanium metal
from reacting with
air/oxygen to reform
titanium dioxide

Pages 72–73   Copper
A
1. c   2. b   3. d   4. d   5. d
B
1. a) true
b) false
c) false
d) false
e) true

2. a) soft
b) copper and zinc
c) copper and tin
d) unreactive
e) lead and tin
C
1. a) chalcopyrite
b) alloy
c) ore
d) brass
2. good thermal insulator
3. a) steel
b) solder
c) amalgam
d) bronze
4. a) it is unreactive/easy to
shape
b) it is unreactive/a good
thermal conductor
c) it is a good electrical
conductor/easy to shape

Pages 74–75   Transition metals
A
1. b   2. d   3. c   4. d   5. b
B
1. a) true
b) true
c) false
d) false
e) true
2. a) amalgam
b) steel
c) bronze
d) brass
C
1. low melting point
2. it is the only non-metal which
conducts electricity
3. a) transition metals
b) iron
4. it is strong
5. it is shiny

Pages 76–77   Noble gases
A
1. d   2. a   3. d   4. d   5. a
B
1. a) electron
b) unreactive
c) increases
d) monatomic
e) colourless
2. a) false
b) true
c) false
d) false
e) true
C
1. a) helium
b) 0
c) monatomic
d) argon

2. a) shared pair of
electrons/covalent bond
b) They already have a full
outer shell of electrons so
they don't need to share
electrons.

3. It has a low density.
4. Hydrogen is flammable.

Pages 78–79   Chemical tests
A
1. d   2. a   3. c   4. b   5. c
B
1. a) carbon dioxide
b) bubbled
c) cloudy/milky
d) lighted
e) squeaky pop
f) oxygen
g) pure
h) glowing
i) relights
j) ammonia
k) damp
l) litmus
m) red
n) blue
o) gas
p) damp
q) bleached
C
1. a) fermentation
b) i)   limewater
ii) limewater goes cloudy
2. damp, red litmus turns blue
3. damp litmus is bleached

Physics
Pages 80–81   Energy
A
1. c   2. c   3. b   4. d   5. d
B
1. (Answers in capitals)

Energy in	Energy changer	Energy out
electrical	bulb	heat and light
chemical	petrol motor	kinetic and heat
electrical	electric motor	kinetic
kinetic	generator	electrical
light	plant leaf	chemical
sound	microphone	electrical
strain potential	catapult	kinetic + gravitational P.E.
electrical	hairdrier	kinetic, heat and sound
chemical	candle	heat and light
chemical	animal	kinetic, heat, chemical
electrical	loud speaker	sound
light	solar cell	electrical
electrical	electric lift	gravitational potential, kinetic
strain potential	bow, clockwork spring	kinetic

C
1. a) 180 J
b) lost to the surroundings
c) eff = 10%
d) eff = 20%
e) to reduce our energy
consumption

Pages 82–83   Generating electricity
A
1. c   2. b   3. d   4. b   5. a

B
1. a)

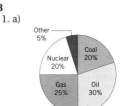

Other 5%
Coal 20%
Nuclear 20%
Gas 25%
Oil 30%

b) solar, tidal, hydroelectric
etc.
2. Coal, oil and **gas** are called
fossil fuels. They are
concentrated sources of
energy. Fossil fuels are formed
from plants and **animals**.
They became covered with

many layers of mud and earth resulting in high pressures and **high** temperatures. Over **millions** of years they changed into fossil fuels. When a fossil fuel is burnt it **releases** energy but releases the gas carbon **dioxide** into the atmosphere. This gas can cause the temperature of the Earth and its atmosphere to increase. This effect is called **the greenhouse effect**. To make fossil fuels last longer we could drive **smaller** cars and turn **down** the heating in our homes.

C
1. a) A fuel is a substance which releases energy when it is burned.
   b) Coal, oil and gas are non-renewable fuels.
   c) Wood, animal dung and methanol are renewable fuels.
   d) chemical energy into heat
   e) heat into kinetic energy
   f) kinetic energy into electrical energy
   g) To reduce the energy lost in the wires of the National Grid during transmission from power station to people's homes.
2. a) No polluting gases are produced.
   b) Risk of nuclear explosion and leaks. Building and decommissioning is very expensive, need to store waste security for many years after use.

Pages 84–85   Renewable sources of energy

A
1. c   2. c   3. b   4. d   5. b

B
1.

Advantage of using this source	Alternative source of energy	Disadvantage of using this source
Using this fuel does not add to the Greenhouse effect	GEOTHERMAL	Obstacle to water traffic
Only low level technology is needed	TIDAL	Large area of land needed for renewal of supply
Energy can be stored until needed	SOLAR	Very high initial construction costs
Useful for isolated island communities	BIOMASS	Few suitable sites
Reliable, available twice a day	WIND	Poor energy capture therefore large area needed
No pollution	HYDROELECTRIC	Possible visual and noise pollution
No pollution and no environmental problems	WAVE	Not useful where there is limited sunshine

C
1. a) gravitational potential energy
   b) kinetic energy
   c) it changes into electrical energy
   d) loss of habitat due to flooding
   e) renewable, no pollution, can be stored till needed
   f) eff = 50%
2. a) The increase in the Earth's temperature caused by more carbon dioxide being in the atmosphere.

b) Reduce the amount of fossil fuels we are burning.

Pages 86–87   Heat transfer – conduction

A
1. a
2. c
3. d
4. c
5. b

B
10% through windows, reduced by installing double glazing
25% through gaps and cracks around doors and windows, reduced by fitting draught excluders
25% through roof, reduced by putting insulation into loft
25% through walls, reduced by having cavity wall insulation
15% through floor, reduced by fitting carpets and underlay

C
1. The correct word order is
   insulator
   conduction
   air
   aluminium
   fibreglass
2. a) double glazing 25 years, loft insulation 2 years, draft excluders 4 years, cavity wall insulation 10 years.
   b) Loft insulation. The savings pay for the insulation in the shortest time.

Pages 88–89   Heat transfer – convection

A
1. d   2. c   3. d   4. c   5. c

B
1. a) day, sea, expands, rises, sea, breezes
   b) During the night the sea is warmer than the land. Air above the sea rises and the cooler air above the land moves out to take its place. It is now an offshore breeze.

C
1. a) Particles must be able to move position if they are to be part of a convection current. Particles in a solid have fixed positions.
   b) Air particles next to the inner wall become warm and rise. Cooler air from the outer wall moves in to take their place.
   c) A convection current is therefore set up which moves heat across the gap.
   d) Air is a good insulator
   e) If cavity wall insulation is put between the walls this stops the movement of the air particles. There is therefore no convection current and no heat loss.
2. a) The fire creates a convection current which carries a lot of heat up the chimney. This energy is lost to the surroundings instead of warming the room.

Pages 90–91   Heat transfer – radiation

A
1. c   2. a   3. b   4. c   5. c

B
1. a) it is reflected
   b) it is absorbed

c) The wax on the back of the dark sheet will melt first and its marble will fall first.
2. Heat radiation from the Sun strikes the silvery surface of the reflector. Most of the radiation is reflected towards the can. Because the can is black it absorbs most of the radiation and the water inside it becomes hot.

C
1. a) the black one
   b) The black car absorbs most of the radiation whilst the white one reflects it.
   c) silver or white
   d) A light coloured card will reflect most of the radiation and so keep the car cooler.
2. a) Plastic is an insulator. Heat loss through the stopper by conduction is therefore very difficult.
   b) Heat is unable to escape across here by conduction or convection as there are no particles.
   c) The far side of the vacuum has a silvered surface which will reflect back any heat crossing the vacuum by radiation.

Pages 92–93   Current, charge and resistance

A
1. b   2. a   3. c   4. d   5. c

B
1. a) 240 Ω
   b) 300 Ω
   c) 150 Ω
   d) 24 Ω
   e) 72 Ω

C
1. a) A resistor whose value can be altered.
   b) By increasing the resistance of the variable resistor.
   c) theatre, cinema etc.
   d) 60 Ω
2. a) A resistor whose resistance changes with light intensity.
   b) If the burglar turns on the light, the resistance of the LDR decreases. Current now flows around the circuit and the buzzer sounds.
   c) 120 Ω

Pages 94–95   Electrical power

A
1. d   2. a   3. b   4. c   5. b

B
1. POWER
   APPLIANCE
   ENERGY
   BILL
   KILOWATT HOUR
   METER
   STANDING CHARGE
   UNIT

C

P	S	T	A	N	D	I	N	G	M
O	C	H	A	R	G	E	H	I	E
W	U	K	I	L	O	W	A	T	T
E	N	H	O	U	R	G	K	J	E
R	I	C	E	F	B	I	L	L	R
B	T	D	E	N	E	R	G	Y	L
A	P	P	L	I	A	N	C	E	M

1. a) 1400 units
   b) £140
   c) This pays for electricity

board equipment and maintenance.
   d) £155
   e) 4
   f) 10 hours
   g) 20 units
   h) £2.00

Pages 96–97   Electric motors

A
1. c   2. c   3. a   4. d   5. d

B
1. a) i)   force up
      ii) force down
      iii) force up
   b) Increase the current in the wire and increase the strength of the magnetic field.

C
1. a) The current in opposite sides of the loop are flowing in opposite directions. As a result one side of the loop is pushed up while the other side is pushed down. The loop therefore rotates.
   b) The wire at the top is still experiencing a force pushing it upwards and the wire at the bottom is still experiencing a force pushing it down.
   c) The split ring changes the direction of the current in the wires each time the loop reaches the vertical position, so the uppermost wire now feels a force pushing it down and the lower wire feels a force pushing it upwards.
   d) Increase the current in the coil, increase the strength of the magnetic field and increase the number of turns on the coil.
   e) Real motors use electromagnets rather than permanent ones. They have several rotating coils not just one. The coils are wrapped around an iron core.
   f) The direction in which the motor is rotating would change.

Pages 98–99   Generators and alternators

A
1. b   2. d   3. d   4. c   5. b

B
1. coil, dynamo, voltage, alternating current, magnetic field, alternator, electricity, generator, induce
2. a) used to generate electricity for a bicycle's light
   b) The magnet no longer moves, so no current is generated and the lights go out.

C
1. a) The needle moves a lot to the left
   b) The needle moves very small amount to the left
   c) The needle does not move
   d) The needle moves to the left
2. a) A current which is continually changing direction.
   b) As the coil is rotated its wires cut through the magnetic field inducing a current in the wires of the coil. Because the wires are

continually changing direction and cutting through the field at different rates the induced current also changes size and direction. It is an alternating current.

c) Rotate the coil faster, use a stronger magnetic field, have more turns on the coil

Pages 100–101   Domestic electricity

A

1. b   2. d   3. b   4. a   5. c

B

1. a) The earth wire is green and yellow and is connected to the top pin. The live wire is brown and connected to the pin on the right. The neutral wire is blue and connected to the pin on the left.
   b) fuse
   c) metal/good conductor/brass
   d) The voltage of the mains supply is much higher than that of a cell or battery and is therefore much more dangerous. It is important therefore that connections in mains circuits are made using insulated plugs.

C

1. a) If the user touches the metal casing of the kettle they may receive an electric shock.
   b) Even if the kettle is faulty the user will not receive an electric shock.
   c) double insulation
   d) Fuses protect people and equipment if a fault develops and too large a current flows in a circuit.
   e) If too large a current passes through the fuse the wire melts making the circuit incomplete and so turning it off.
   f) Any two of: 1 A, 3 A and 13 A
   g) It's a special kind of fuse that can usually be reset by pushing a button.
   h) When using an electric lawn mover or hedge trimmers.

2. a) Current which only flows in one direction.
   b) Current which is continuously changing direction
   c) Cell or battery.
   d) The mains or household electricity.

Pages 102–103   Waves

A

1. c   2. a   3. c   4. a   5. d

B

1. REFRACTED
   REFLECTED
   LIGHT
   VIBRATION
   FREQUENCY
   SOUND
   WAVELENGTH
   HERTZ
   SEISMIC
   SPEED
   TRANSVERSE
   AMPLITUDE
   LONGITUDINAL

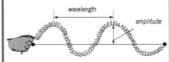

C

1. a)

   b) vibrations are at right angles to the direction the wave is moving
   c) light/surface water wave
   d) vibrations are along the direction in which the wave is moving.
   e) sound wave
   f) 200 m/s

2. a) Earthquakes
   b) P-waves and S-waves
   c) Any of: P-waves are longitudinal waves. S-waves are transverse waves. P-waves travel faster than S-waves. P waves can travel through solids and liquids. S-waves cannot travel through liquids.
   d) Scientists have used these waves to investigate the internal structure of the Earth.

Pages 104–105   Electromagnetic spectrum 1

A

1. c   2. d   3. c   4. a   5. c

B

1. a) radio waves, microwaves, ultraviolet, infrared, X-rays, gamma rays, visible light
   b) transverse, refraction, transmission, reflection, absorption,

C

1. a) A – visible light, B – X-rays
   b) radio waves
   c) visible light
   d) gamma waves and X-rays

2. a) The wave passes through the object.
   b) The wave bounces off the object.
   c) The wave is taken in by the object.
   d) How long the object is exposed, the nature of the objects surface and the material the object is made from.
   e) Warm the skin. Tan or burn the skin. Premature aging of the skin. Possibly cause cancer.

Pages 106–107   Electromagnetic spectrum 2

A

1. d   2. b   3. b   4. a   5. c

B

a) false
b) true
c) false
d) true
e) false
f) true
g) true
h) true
i) false
j) true

C

a) A layer of charged particles high above the Earth.
b)

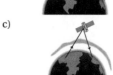

c)

d) Analogue signals can become distorted as they travel. Digital do not.

2. a) infrared
   b) reflection
   c) Infrared waves have fairly low penetrating powers and so are unable to pass through the wall.

Pages 108–109   Nuclear radiation

A

1. d   2. a   3. c   4. b   5. b

B

1. a) false
   b) true
   c) false
   d) false
   e) false
   f) true
   g) false
   h) true
   i) true
   j) false

C

1. a) The nucleus
   b) The upper radiation is gamma. The lower is beta radiation
   c) An ion is an atom which has lost or gained electrons and so is charged.
   d) Radiation collides with atoms knocking off electrons.
   e) Alpha, beta and then gamma

2.

3.

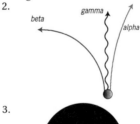

Pages 110–111   Uses of radioactivity

A

1. b   2. d   3. c   4. c   5. a

B

1. more radiation reaches the detector and the pressure applied by the rollers is decreased.
   less radiation reaches the detector and the pressure applied by the rollers is increased.
   thickness/quality of the metal sheet is correct.
   paper.

2. a) The presence and growth of bacteria.
   b) If food is exposed to gamma radiation the bacteria are killed.

c) sterilising surgical instruments

C

1. a) gamma
   b) The cancerous cells are killed.
   c) The dose of radiation received at B and C is too small to damage/kill cells.

2. a) oil, gas
   b) a leak in the pipe (possibly a blockage)
   c) increase in count rate shown by detector (sudden fall in detection of tracer cannot penetrate blockage)
   d) Alpha and beta radiations would be unable to pass through soil/earth above pipe.
   e) Avoids the need to dig up whole section of pipes to find leak/blockage.

Pages 112–113   The Earth and our solar system

A

1. b   2. b   3. a   4. d   5. a

B

1. SATELLITE
   SOLAR SYSTEM
   SATURN
   COMET
   URANUS
   PLUTO
   STAR
   ASTEROID
   MOON
   MARS
   SUN
   EARTH
   JUPITER
   ORBIT

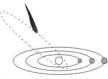

C

1. a) Large rock-like piece of ice that orbits the Sun.
   b) Large piece of rock that orbits the Sun.
   c) In a belt between Mars and Jupiter.
   d) The orbit of a comet takes it close to the Sun and to the far reaches of the solar system. ie distance from the Sun varies enormously.
   e) gravitational forces
   f)

   g) When it is closest to the Sun.

2. a) A natural satellite of a planet is a moon. An artificial satellite is a man-made object which orbits a planet.
   b) weather satellite, communications satellite etc.

Pages 114–115 Stars and the universe

A

1. a 2. d 3. d 4. b 5. a

B

1. a) Gases are pulled together by gravitational forces. These forces cause the gases to be compressed and their temperatures increase.
   b) Temperature sets off nuclear reactions.
   c) Large amounts of energy as heat and light. The star is formed.
   d) Dust and gases may gather to form planets and moons.

C

1. a) gravitational forces
   b) nuclear reactions
   c) energy

d) the star is in its main stable period
e) an exploding star
f) a very dense neutron star or a black hole

2. a) A theory which suggests that the universe began with all the matter in one place. This matter then exploded in all directions.
   b) All the galaxies we can see are moving away from us and the further away they are the faster they are moving.
   c) The Universe will continue to expand for ever or the expansion will gradually slow, stop and then reverse pulling all the matter back into one place.

3. a) A very large group of stars (billions of stars).
   b) The Milky Way.

Pages 116–117 Exploring space

A

1. b 2. d 3. c 4. c 5. d

B

1. a) false
   b) false
   c) true
   d) false
   e) false
   f) true
   g) false
   h) true
   i) false
   j) true
   k) false
   l) true
   m) false
   n) true
   o) true

C

1. a) To avoid distortions caused by the Earth's atmosphere.

   b) Infrared, ultraviolet, visible light, radio waves.

2. a) A flyby probe investigates a planet/moon without landing. A lander lands on the surface of a moon or planet.
   b) Information about soil, atmosphere, gravitational or magnetic fields etc.
   c) Cheaper and less hazardous. No need to provide food, oxygen etc for human onboard
   d) Muscle wastage, calcium depletion.
   e) Daily exercise, artificial gravity.
   f) Air/oxygen supply, food and water, fuel for return journey, radiation shield.

ACKNOWLEDGEMENTS

The author and publisher are grateful to the copyright holders for permission to use quoted materials and photographs.

Letts and Lonsdale
4 Grosvenor Place
London SW1X 7DL

School orders: 015395 64910
School enquiries: 015395 65921
Parent and student enquiries: 015395 64913
Email: enquiries@lettsandlonsdale.co.uk
Website: www.lettsandlonsdale.com

First published 2006

02/220808

Text, design and illustration © 2006 Letts Educational Ltd

British Library Cataloguing in Publication Data. A CIP record of this book is available from the British Library.

ISBN: 9781843156697

Book concept and development: Helen Jacobs, Publishing Director

Letts editorial team: Catherine Dakin

Series Editor: Brian Arnold

Authors: Brian Arnold, Elaine Gill and Emma Poole

Cover design: Angela English

Inside concept design: Starfish Design

Text design, layout and editorial: MCS Publishing Services

Letts and Lonsdale make every effort to ensure that paper used in our books is made from wood pulp obtained from well-managed forests.

C

1 The Earth's lithosphere is split into about twelve plates. Scientists believe that these plates are slowly moving. The diagram below shows South America and Africa. It is thought that South America and Africa were once joined together.

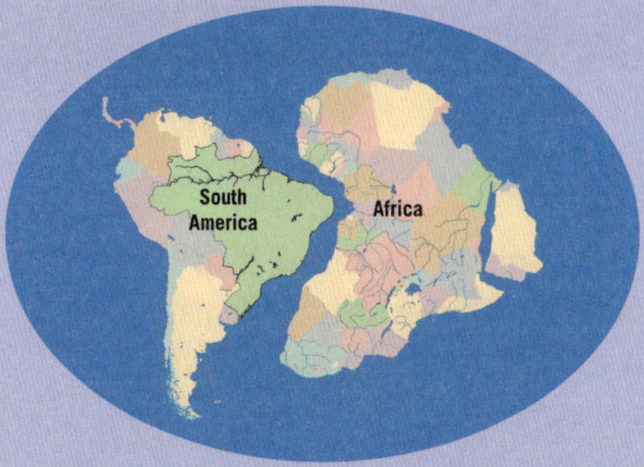

a) What is the Earth's lithosphere? (1 mark)

....

....

....

....

....

....

b) Give two pieces of evidence that suggest that South America and Africa could once have been joined together. (2 marks)

....

....

....

....

....

....

c) Why do the Earth's tectonic plates move? (1 mark)

....

....

....

....

....

d) How fast are the Earth's plates moving? Circle one answer. (1 mark)

a few mms per year

a few cms per year

a few ms per year

a few kms per year

e) The Earth has a layered structure. Which option best describes the state of the inner and outer core? Circle one answer. (1 mark)

inner core	outer core
solid	solid
liquid	solid
liquid	liquid
solid	liquid

f) Which layer of the Earth is mainly made of silicon, oxygen and aluminium? (1 mark)

crust

mantle

outer core

inner core

How well did you do? ✗ 0-8 Try again 9-13 Getting there 14-17 Good work 18-22 Excellent! ✓

Consequences of plate tectonics

A

1 Where is the San Andreas Fault? **(1 mark)**

a) Kansas
b) Florida
c) Arizona
d) California

2 The movement of tectonic plates causes many problems. Which of these statements is NOT true? **(1 mark)**

a) The problems are worst along plate boundaries.
b) The problems are worst at the centre of plates.
c) The problems include earthquakes.
d) The problems include volcanoes.

3 Where is the Andes mountain range? **(1 mark)**

a) West coast of South America
b) East coast of South America
c) West coast of Africa
d) East coast of India

4 What can be caused when an earthquake occurs under the ocean? **(1 mark)**

a) hot water
b) a tsunami
c) new granite rocks
d) they cannot happen under water

5 Oceanic plates are denser than continental plates. Which elements are present in the highest amounts in oceanic rocks? **(1 mark)**

a) oxygen and aluminium
b) iron and oxygen
c) magnesium and iron
d) aluminium and iron

B

1 Label the diagram to show what can happen when a continental plate and an oceanic plate collide.

Use the labels shown in the box. **(5 marks)**

| melting | fold mountains | oceanic plate |
| continental plate | volcano | |

2 Complete the following passage. **(7 marks)**

forces   San Andreas   boundaries   earthquake   slide   plates   stuck

Earthquakes often occur along plate .... They happen when the plates .... past each other. A famous earthquake zone is along the .... Fault in California. The .... in this area have been broken into a complicated pattern. As the plates slip past each other they often get .... together. The .... on the plates gradually build up, until eventually the plates move and the strain is released as an ....

C

1 This diagram shows the layered structure of the Earth.

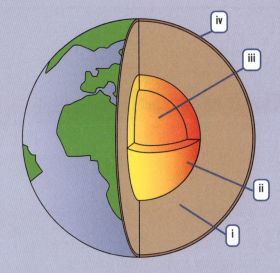

a) Add the missing labels to complete the diagram. **(4 marks)**

i) ....

ii) ....

iii) ....

iv) ....

b) What is the name scientists give to the crust and the upper part of the mantle?

(1 mark)

....

....

c) The crust and upper mantle are split into a number of moving plates. What type of rock is formed when plates are stressed at a plate boundary? Circle one answer. **(1 mark)**

metamorphic

sedimentary

intrusive igneous

extrusive igneous

d) Sometimes an oceanic plate and a continental plate collide. Why is oceanic plate forced beneath continental plate?

(1 mark)

....

....

....

e) When oceanic plate and continental plates collide mountain ranges can be formed. Where, in the world, is this process forming mountain ranges today? Circle one answer.

(1 mark)

Iceland

West coast of South America

California

West coast of Africa

f) Give an example of a natural disaster which is associated with problems at plate boundaries. **(1 mark)**

....

....

....

g) Why can we not predict exactly when an earthquake will occur? **(1 mark)**

....

....

....

How well did you do? ✗ **0-11** Try again   **12-16** Getting there   **17-22** Good work   **23-27** Excellent! ✓

Extraction of iron

A

1 Which of these metals can be found on its own in nature? *(1 mark)*

a) potassium
b) iron
c) gold
d) calcium

2 Which of these substances is NOT a solid added to the blast furnace? *(1 mark)*

a) iron ore
b) carbon dioxide
c) coke
d) limestone

3 What is the name given to the chemical reaction in which oxygen is removed from iron oxide? *(1 mark)*

a) neutralisation
b) reduction
c) oxidation
d) endothermic

4 What is the name of the main ore of iron? *(1 mark)*

a) iron sulphide
b) magnetite
c) galena
d) haematite

5 What is the name of the substance that mainly reduces iron oxide to iron in the blast furnace? *(1 mark)*

a) carbon
b) coke
c) carbon dioxide
d) carbon monoxide

B

1 Complete the following passage. *(8 marks)*

slag   oxide   limestone   haematite   furnace   silicon dioxide   density   slag

The main ore of iron is called .... It contains iron .... Haematite often contains some impurities. The main impurity in haematite is normally .... , which is often known as silica. When .... is added to the blast .... it reacts with the silica to form a molten substance called .... The .... has a lower .... and floats on top of the molten iron.

2 True or false? *(5 marks)*

	true	false
a) All ores contain at least 50% metal.		
b) Coke is a good source of the element carbon.		
c) The less reactive a metal is, the harder it is to remove form its compound.		
d) In the blast furnace carbon monoxide is oxidised to carbon dioxide.		
e) Molten iron has a lower density so it sinks to the bottom of the blast furnace where it can be removed.		

C

1

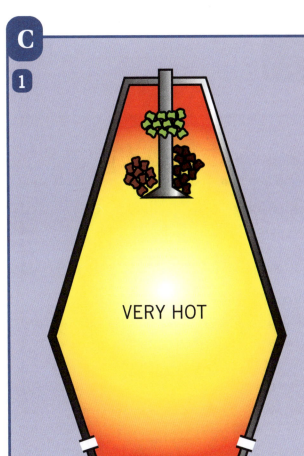

VERY HOT

This diagram shows a blast furnace which is used to extract iron from iron ore.

a) Three solid raw materials are added to the blast furnace. One of the solid raw materials is iron ore. Name the other two.   **(2 marks)**

....

....

....

b) The main chemical compound in iron ore is iron oxide. What is the main ore of iron called?   **(1 mark)**

....

....

....

c) What is the other raw material added to the blast furnace?   **(1 mark)**

....

....

....

d) Complete this equation to show the reaction between carbon monoxide and iron oxide.   **(1 mark)**

Carbon monoxide **+** iron oxide → iron + ...

e) In the blast furnace iron is extracted from iron ore. Which of these words best describes what happens to the iron in iron oxide? Tick one box.   **(1 mark)**

electrolysis ☐
reduction ☐
oxidation ☐
neutralisation ☐

f) In the blast furnace what does the limestone do?   **(1 mark)**

....

....

....

2 In the blast furnace these reactions take place.

Carbon + oxygen → carbon dioxide

Carbon dioxide + carbon → carbon monoxide

Which substance is oxidised in both reactions?   **(1 mark)**

....

....

....

Iron and steel

A

1 Which of these metals is NOT found in stainless steel? **(1 mark)**

a) iron ☐
b) titanium ☐
c) chromium ☐
d) nickel ☐

2 Which of these properties is NOT true of high carbon steel? **(1 mark)**

a) hard ☐
b) strong ☐
c) easy to shape ☐
d) brittle ☐

3 Which of these elements can be added to pure iron to make steel? **(1 mark)**

a) sulphur ☐
b) carbon ☐
c) oxygen ☐
d) silicon ☐

4 Which of these ways would NOT stop iron from rusting? **(1 mark)**

a) painting ☐
b) cover in oil ☐
c) cover in plastic ☐
d) cover in water ☐

5 Roughly how much carbon is found in cast iron? **(1 mark)**

a) 1% ☐
b) 4% ☐
c) 10% ☐
d) 40% ☐

B

1 Complete these sentences by crossing out the incorrect word/phrase. **(5 marks)**

a) Coating iron in plastic stops it rusting because it stops oxygen and water/oxygen and salt reaching the iron.
b) Iron can be protected from rusting by placing it in contact with a more/less reactive metal.
c) Iron can be mixed with the metallic/ non-metallic element carbon to form the alloy steel.
d) Cast iron is softer/harder than wrought iron.
e) Steel is harder than iron because it consists of atoms of different sizes/ the same size.

2 Complete the following passages. **(11 marks)**

Iron from the blast furnace contains high levels of the element .... If this iron is cooled down until it solidifies it forms .... iron. Cast iron is hard, strong and does not .... However, cast iron has a notable disadvantage; it is very .... Cast iron can be used to make objects like ....

Wrought iron is made by removing .... from cast iron. Wrought iron is much .... than cast iron. It is also much easier to .... In wrought iron, the iron .... form a very regular arrangement. This means that the layers of atoms are able to .... easily over each other. Wrought iron can be used to make objects like ....

1 a) What two substances must be present for iron to rust? **(1 mark)**

....

....

....

....

b) Explain how coating an iron fence in plastic can stop the iron from rusting. **(1 mark)**

....

....

....

....

2 This diagram shows the structure of cast iron.

silicon atoms

carbon atoms

iron atoms

a) Complete the box below to show the structure of wrought iron. Label the atoms in your diagram. **(2 marks)**

b) Would you expect cast iron or wrought iron to be harder? Explain your answer. **(1 mark)**

....

....

....

....

....

c) Iron can be made into steel by adding carbon and other metals such as chromium. What is a mixture of metals called? **(1 mark)**

....

....

....

....

....

d) Steel can be used to make hammers. Which of these properties should a hammer have? Tick one answer. **(1 mark)**

soft ☐
hard to shape ☐
brittle ☐
strong ☐

e) How does increasing the amount of carbon in steel affect how easy the carbon is to shape? **(1 mark)**

....

....

....

....

....

Aluminium

A

1 Why are drinks cans made from aluminium? **(1 mark)**

a) Aluminium is not reactive. ☐
b) Aluminium does not react with acids. ☐
c) Aluminium does not react with oxygen. ☐
d) Aluminium reacts with oxygen to form a layer of aluminium oxide which prevents any further reaction. ☐

2 What is the name of the compound found in bauxite? **(1 mark)**

a) iron oxide ☐
b) aluminium iodide ☐
c) aluminium oxide ☐
d) aluminium oxygen ☐

3 Which of these metals is the most reactive? **(1 mark)**

a) zinc ☐
b) iron ☐
c) copper ☐
d) aluminium ☐

4 What is the main ore of aluminium? **(1 mark)**

a) cryolite ☐
b) bauxite ☐
c) dolomite ☐
d) magnetite ☐

5 Why is aluminium extracted from its ore by electrolysis? **(1 mark)**

a) Aluminium is more reactive than carbon. ☐
b) Aluminium is less reactive than carbon. ☐
c) Electrolysis is very cheap. ☐
d) Electrolysis is easy to do. ☐

B

1 The diagram shows how aluminium can be extracted from its ore by electrolysis. Label the diagram using the labels shown in the box.

(4 marks)

positive ion	negative ion
positive electrode	negative electrode

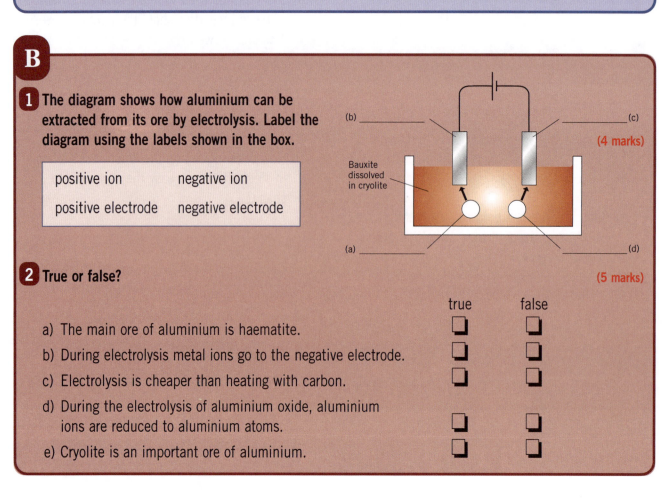

(b) _____

(c) _____

Bauxite dissolved in cryolite

(a) _____

(d) _____

2 True or false?

(5 marks)

	true	false
a) The main ore of aluminium is haematite.	☐	☐
b) During electrolysis metal ions go to the negative electrode.	☐	☐
c) Electrolysis is cheaper than heating with carbon.	☐	☐
d) During the electrolysis of aluminium oxide, aluminium ions are reduced to aluminium atoms.	☐	☐
e) Cryolite is an important ore of aluminium.	☐	☐

C

1 This diagram shows how the metal aluminium can be extracted from aluminium oxide.

Power supply

a) What is the name given to this process?
(1 mark)

....

....

....

b) Name the main ore of aluminium. (1 mark)

....

....

....

c) Name the other ore of aluminium that is also used in the extraction of aluminium.
(1 mark)

....

....

....

d) Why is the other ore of aluminium used?
(1 mark)

....

....

....

....

e) During the electrolysis of aluminium oxide, the aluminium ions move. Which of these words best describes what happens to aluminium ions during electrolysis? Tick one box. (1 mark)

displacement ☐
reduction ☐
oxidation ☐
neutralisation ☐

f) During the electrolysis of aluminium oxide, the oxide ions also move. Which electrode do these ions move towards? (1 mark)

....

....

....

g) On the diagram above, label the positive electrode. (1 mark)

h) What material is the positive electrode made from? (1 mark)

....

....

....

i) Why must the positive electrode be periodically replaced? (1 mark)

....

....

....

How well did you do? ✗ **0-9** Try again   **10-14** Getting there   **15-18** Good work   **19-23** Excellent! ✓

Titanium

A

1 Which of these metals is the most reactive? **(1 mark)**

a) titanium
b) copper
c) iron
d) gold

2 Titanium is a useful metal. Which of these properties does titanium NOT have? **(1 mark)**

a) low density
b) difficult to shape
c) very high melting point
d) resistant to corrosion

3 What is the main ore of titanium? **(1 mark)**

a) rutile
b) haematite
c) chalcopyrite
d) chalcosine

4 Why does titanium appear to be less reactive than its position in the reactivity series suggests? **(1 mark)**

a) Titanium forms a layer of titanium sulphide which prevents any further reaction.
b) Titanium is very resistant to corrosion.
c) Titanium forms a layer of aluminium oxide which prevents any further reaction.
d) Titanium forms a layer of titanium oxide which prevents any further reaction.

5 What is a mixture of metals called? **(1 mark)**

a) compound
b) mixture
c) emulsion
d) alloy

B

1 True or false? **(5 marks)**

	true	false
a) Titanium is extracted from its ore.	☐	☐
b) Titanium is less reactive than iron.	☐	☐
c) The main ore of titanium, rutile is very resistant to corrosion.	☐	☐
d) Magnesium is more reactive than titanium.	☐	☐
e) Titanium dioxide has the formula TiO.	☐	☐

2 Complete these sentences by crossing out the incorrect word/phrase. **(5 marks)**

a) Pure titanium is too soft/hard for many uses.
b) Nitinol is an alloy of nickel and copper/nickel and titanium.
c) Titanium metals appear to be more reactive/less reactive than they really are because titanium forms a layer of titanium oxide which prevents any further reaction.
d) When shape memory alloys are heated they change colour/return to their original shape.
e) In the extraction of titanium the titanium is extracted by displacement by molten magnesium. Magnesium is more reactive/less reactive than titanium.

C

1 This question is about copper and titanium metals.

Use the words below to complete the table. **(4 marks)**

nitinol   magnesium   alloy   rutile

Name	Description
a)	the main ore of titanium
b)	a mixture of metals
c)	the metal used to displace titanium from titanium chloride
d)	a smart alloy

2 Titanium is a very useful metal.

a) Which of these properties does titanium **NOT** have? Tick one box. **(1 mark)**

high density ☐

high melting point ☐

easy to shape ☐

high resistance to corrosion ☐

```
┌─────────────────────┐
│   titanium oxide    │
└─────────────────────┘
     ↓ stage ①
┌─────────────────────┐
│  titanium chloride  │
└─────────────────────┘
     ↓ stage ②
┌─────────────────────┐
│      titanium       │
└─────────────────────┘
```

This flow diagram shows stages in the manufacture of titanium metal from titanium oxide. In the first stage titanium oxide is converted to titanium chloride. In the second stage titanium chloride is reacted with molten magnesium.

b) Give the word equation for the reaction between magnesium and titanium chloride. **(1 mark)**

....

....

....

....

....

....

c) Why could this reaction be described as a displacement reaction? **(1 mark)**

....

....

....

....

....

....

d) Why must this reaction be carried out under a vacuum? **(1 mark)**

....

....

....

....

....

....

Copper

A

1 **How long have people used copper?** (1 mark)

a) since the mid-twentieth century ☐
b) since Roman times ☐
c) since ancient times ☐
d) since 1800. ☐

2 **Copper is a useful metal. In which of these applications is copper NOT used in large quantities?** (1 mark)

a) water pipes ☐
b) aircraft bodies ☐
c) saucepans ☐
d) electrical wiring ☐

3 **Which of these minerals is an ore of copper?** (1 mark)

a) rutile ☐
b) haematite ☐
c) bauxite ☐
d) chalcosine ☐

4 **Copper is a useful metal. Which of these properties does copper NOT have?** (1 mark)

a) good thermal conductor ☐
b) resistant to corrosion ☐
c) unreactive ☐
d) poor electrical conductor ☐

5 **What is a mixture of metals called?** (1 mark)

a) compound ☐
b) mixture ☐
c) emulsion ☐
d) alloy ☐

B

1 **True or false?** (5 marks)

	true	false
a) Copper is a good thermal conductor.	☐	☐
b) There are concerns that cooking acidic foods like rhubarb in copper saucepans could cause long-term health problems.	☐	☐
c) Copper is purified by filtering the hot metal to remove impurities.	☐	☐
d) Gold is more reactive than copper.	☐	☐
e) When copper is extracted from copper oxide the copper is reduced.	☐	☐

2 **Complete these sentences by crossing out the incorrect word/phrase.** (5 marks)

a) Pure copper is too soft/hard for many uses.
b) Brass is an alloy of copper and zinc/tin.
c) Bronze is an alloy of copper and zinc/tin.
d) Copper is a very reactive/unreactive metal.
e) Solder is made from lead and tin/lead and mercury.

1 This question is about copper.

Use the words below to complete
the table. (4 marks)

ore
chalcopyrite
alloy
brass

Name	Description
a)	an ore of copper
b)	a mixture of metals
c)	a rock which contains a metal
d)	made from copper and zinc

2 Copper is a very useful metal. It can be used to make objects like saucepans and electrical wires.

Which of these properties
does copper NOT have? Tick one box. (1 mark)

good thermal insulator ☐

good electrical conductor ☐

easy to shape ☐

high resistance to corrosion ☐

3 This question is about alloys.

Use the words below to complete
the table. (4 marks)

bronze
steel
solder
amalgam

Name	Description
a)	is made from iron
b)	is made from lead and tin
c)	contains mercury
d)	made from copper and tin

4 Copper has some very useful properties. It is a very widely used metal.

a) Why is copper used to make water pipes?
(1 mark)

....

....

....

....

b) Why is copper used to make saucepans?
(1 mark)

....

....

c) Why is copper used to make electrical wire?
(1 mark)

....

....

Transition metals

A

1 Which of these transition metals rusts? (1 mark)

a) copper ☐
b) iron ☐
c) nickel ☐
d) platinum ☐

2 Which of these properties do you NOT expect of a metal? (1 mark)

a) good electrical conductor ☐
b) good thermal conductor ☐
c) can be hammered into shape ☐
d) it is a gas at room temperature ☐

3 Which of these properties do you NOT expect of a transition metal? (1 mark)

a) hard ☐
b) strong ☐
c) forms white compounds ☐
d) tough ☐

4 What is a mixture of metals called? (1 mark)

a) emulsion ☐
b) compound ☐
c) oxide ☐
d) alloy ☐

5 Which of these examples is NOT a common alloy? (1 mark)

a) steel ☐
b) iron ☐
c) brass ☐
d) amalgam ☐

B

1 True or false? (5 marks)

	true	false
a) A catalyst increases the rate of a chemical reaction.	☐	☐
b) Transition metals are often good catalysts.	☐	☐
c) Transition metals form white compounds.	☐	☐
d) Copper is used in the Haber process.	☐	☐
e) Nickel is used in the manufacture of margarine.	☐	☐

2 Complete the table to show the names of some common alloys. Use the names below. (4 marks)

steel   brass   amalgam   bronze

Name of alloy	What is the alloy made of?
a)	mainly mercury
b)	iron, carbon and other metals like chromium
c)	copper and tin
d)	copper and zinc

C

1 Which of these properties is NOT typical of a transition metal?
Tick one box. (1 mark)

hard wearing ❏

good thermal conductor ❏

strong ❏

low melting point ❏

2 Graphite is a form of carbon. It is a dark grey solid substance which is soft, brittle and conducts electricity.

Which property of graphite makes it a very unusual non-metal? (1 mark)

....

....

....

3 The diagram below shows a section of the periodic table.

		'middle block'					
	Ti	Cr	Fe	Ni	Cu	Zn	

a) What is the name given to the area of the periodic table described in the diagram as the 'middle block'? (1 mark)

....

....

b) What is the name of the element which has the symbol Fe? (1 mark)

....

....

4 Which of these properties makes iron a good metal to make cars from?
Tick one box. (1 mark)

it rusts ❏

it is heavy ❏

it is strong ❏

5 Which of these properties makes nickel a good metal to make coins from?
Tick one box. (1 mark)

it is shiny ❏

it is a good catalyst ❏

it is a good electrical conductor ❏

Noble gases

A

1 **What is the most unreactive group in the periodic table?** (1 mark)

a) Group 1 ☐
b) Group 2 ☐
c) Group 7 ☐
d) Group 0 ☐

2 **What does monatomic mean?** (1 mark)

a) individual atoms ☐
b) one colour ☐
c) one outer electron ☐
d) they are colourless ☐

3 **How many electrons do noble gas atoms have to gain to get a full outer shell?** (1 mark)

a) 1 ☐
b) 2 ☐
c) 7 ☐
d) 0 ☐

4 **Which noble gas is used in light bulbs?** (1 mark)

a) helium ☐
b) neon ☐
c) krypton ☐
d) argon ☐

5 **Which noble gas is used in air balloons?** (1 mark)

a) helium ☐
b) neon ☐
c) krypton ☐
d) argon ☐

B

1 **Complete the following sentences by crossing out the incorrect word/phrase.** (5 marks)

a) The noble gases all have a full outer electron/proton shell.
b) Noble gases are very reactive/unreactive.
c) The size of noble gas atoms increases/decreases down the group.
d) Noble gases are monatomic/diatomic.
e) Noble gases are pastel coloured/colourless.

2 **True or false?** (5 marks)

	true	false
a) Noble gases have no practical uses.	☐	☐
b) Noble gases already have a full and stable outer shell of electrons.	☐	☐
c) Pairs of noble gas atoms join together to form molecules such as He_2.	☐	☐
d) The noble gases range in colour from pale blue to pale pink.	☐	☐
e) Down the group the noble gas atoms get larger.	☐	☐

C

1 This question is about noble gases.

Use the words below to complete the table. (4 marks)

argon
monatomic
helium
0

Name	Description
a)	used in balloons
b)	the group of the periodic table where the noble gases are found
c)	single atoms
d)	used in filament light bulbs

2 This diagram shows how two atoms can join together to form a hydrogen molecule.

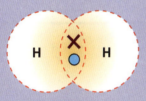

The dots and crosses represent electrons. In the first electron shell there is room for up to two electrons.

a) How are the two hydrogen atoms held together? (1 mark)

....

....

....

This diagram shows a helium atom.

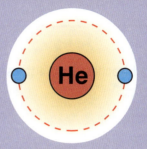

b) Why don't helium atoms join together to form helium molecules? (1 mark)

....

....

....

3 Helium can be used in balloons.

Why do we use helium in party balloons? (1 mark)

....

....

....

4 Why is it better to use helium than hydrogen in a party balloon? (1 mark)

....

....

....

Chemical tests

A

1 Which gas bleaches damp litmus? (1 mark)

a) oxygen ☐
b) carbon dioxide ☐
c) hydrogen ☐
d) chlorine ☐

2 Which gas relights a glowing splint? (1 mark)

a) oxygen ☐
b) carbon dioxide ☐
c) ammonia ☐
d) chlorine ☐

3 Which gas burns with a squeaky pop? (1 mark)

a) oxygen ☐
b) carbon dioxide ☐
c) hydrogen ☐
d) chlorine ☐

4 Which gas turns limewater cloudy? (1 mark)

a) oxygen ☐
b) carbon dioxide ☐
c) hydrogen ☐
d) chlorine ☐

5 Which gas turns damp red litmus paper blue? (1 mark)

a) oxygen ☐
b) carbon dioxide ☐
c) ammonia ☐
d) chlorine ☐

B

1 Complete the following sentences. (17 marks)

litmus	glowing	bleached	squeaky pop	oxygen	carbon dioxide
pure	damp	bubbled	relights	cloudy/milky	ammonia
red	lighted	blue	damp	gas	

Gas tests

Limewater is used to test for the gas a) .... The gas is b) .... through limewater. If the limewater turns c) .... the gas is carbon dioxide.

The gas hydrogen is tested for using a d) .... splint. If hydrogen is present it will produce a e) ....

The gas f) .... is needed for things to burn. Things burn more brightly in g) .... oxygen than they do in air. If a h) .... splint is placed in a test tube containing oxygen, the splint i) ....

The gas j) .... is tested for using k) .... , red l) .... paper. If ammonia is present the litmus paper changes colour from m) .... to n) ....

The o) .... chlorine is tested for using p) .... litmus paper. If chlorine is present the litmus paper is q) ....

C

1 This diagram shows the equipment that is used to turn sugar into alcohol and carbon dioxide.

flask

water and sugar

test tube (A)

bubbles

a) What is the name of this chemical reaction? **(1 mark)**

....

....

....

b) i) What should be added to test tube A to test that carbon dioxide has been made? **(1 mark)**

....

....

....

ii) What would you expect to see to solution in test tube A if carbon dioxide is being made? **(1 mark)**

....

....

....

2 The reaction between ammonium nitrate and sodium hydroxide produces sodium nitrate, ammonia and water.

This reaction can be summarised by the word equation

$$\text{Ammonium nitrate} + \text{sodium hydroxide} \rightarrow \text{sodium nitrate} + \text{ammonia} + \text{water}$$

How could you tell that ammonia gas has been made? What would you do and what would you expect to see? **(2 marks)**

....

....

....

3 A student carries out an experiment on an unknown compound. The compound is a bright pink colour. This means that the compound contains the transition metal cobalt. The student believes that the compound could be cobalt chloride. When cobalt chloride is heated fiercely, it gives off the gas chlorine. Explain how the student should test the gas given off when the compound is heated to see if it is chlorine. **(2 marks)**

....

....

Energy

A

1 Which of the following is not a form of energy? (1 mark)

a) chemical ☐
b) electrical ☐
c) power ☐
d) sound ☐

2 What kind of energy is stored in a stretched elastic band? (1 mark)

a) sound ☐
b) chemical ☐
c) strain potential energy ☐
d) gravitational potential energy ☐

3 What device could be used to change electrical energy into sound energy? (1 mark)

a) microphone ☐
b) radio ☐
c) megaphone ☐
d) trombone ☐

4 As water falls down a waterfall it gains (1 mark)

a) potential energy ☐
b) strain energy ☐
c) chemical energy ☐
d) kinetic energy ☐

5 Which of the following statements is not true? (1 mark)

a) Energy cannot be created. ☐
b) Energy cannot be destroyed. ☐
c) Energy can be stored. ☐
d) Energy can never be seen or heard. ☐

B

1 Fill in the empty spaces in the table below. The first one has been done for you. (20 marks)

Energy in	Energy changer	Energy out
electrical	bulb	heat and light
	petrol motor	
	electric motor	
	generator	
	plant leaf	
	microphone	
	catapult	
	hairdrier	
chemical		heat and light
chemical		kinetic, heat, chemical
electrical		sound
light		electrical
electrical		gravitational potential
strain potential		kinetic

C

1 The diagram below shows the energy changes that take place in a bulb.

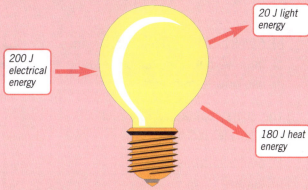

20 J light energy

200 J electrical energy

180 J heat energy

a) How much energy is wasted during this transfer? **(1 mark)**

....

....

....

....

b) Where does this energy go? **(1 mark)**

....

....

....

....

c) Assuming that the useful energy required from the bulb is light energy, calculate the efficiency of the bulb. **(3 marks)**

....

....

....

....

The diagram below shows a compact fluorescent light.

15W

d) If the compact fluorescent light changes 200 J of electrical energy into 40 J of light energy and 160 J of heat, calculate its efficiency. **(3 marks)**

....

....

....

....

....

....

....

....

e) Why is it important that we all use devices which are more efficient? **(1 mark)**

....

....

....

....

Generating electricity

A

1 Which of the following is not a fossil fuel? **(1 mark)**

a) oil ☐
b) gas ☐
c) wood ☐
d) coal ☐

2 Approximately how long does it take for fossil fuels to form? **(1 mark)**

a) a few years ☐
b) a few million years ☐
c) a few centuries ☐
d) less than a year ☐

3 Which of the following will not slow down the rate at which we are using up our supplies of fossil fuels? **(1 mark)**

a) use alternative sources of energy ☐
b) develop more efficient car engines ☐
c) improve home insulation ☐
d) use private transport ☐

4 The energy contained in a fuel can be released by **(1 mark)**

a) photosynthesis ☐
b) burning ☐
c) electrolysis ☐
d) electroplating ☐

5 In a power station the kinetic energy of the turbines is changed into **(1 mark)**

a) electrical energy ☐
b) chemical energy ☐
c) potential energy ☐
d) heat ☐

B

1 The list below gives details of the different sources of energy used in the UK in a year.

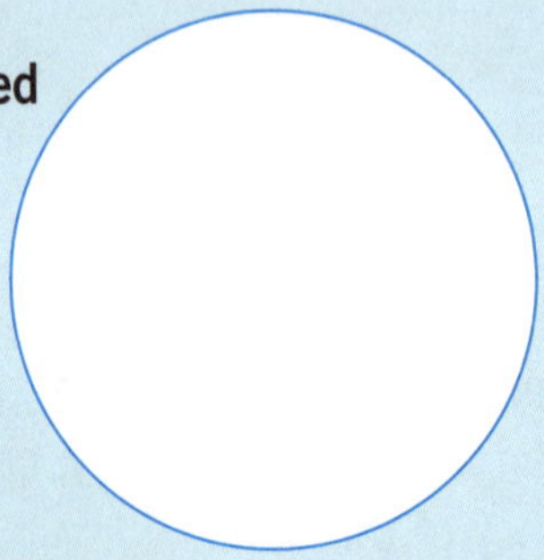

Coal	20%
Oil	30%
Gas	25%
Nuclear	20%
Other	5%

a) Using the circle drawn above, represent these figures as a pie chart. **(4 marks)**

b) Suggest one energy source which might be included in the 5% sector of your pie chart. **(1 mark)**

2 The following account of fossil fuels contains lots of errors. Re-write the account correcting any errors you find. **(10 marks)**

Coal, oil and wood are called fossil fuels. They are dilute sources of energy. Fossil fuels are formed from plants and rocks. They became covered with many layers of mud and earth resulting in high pressures and low temperatures. Over thousands of years they changed into fossil fuels.

When a fossil fuel is burnt it takes in energy but releases the gas carbon monoxide into the atmosphere. This gas can cause the temperature of the Earth and its atmosphere to increase. This effect is called acid rain. To make fossil fuels last longer we could drive bigger cars and turn up the heating in our homes.

C

1 a) What is a fuel? **(1 mark)**

....

....

b) Name two non-renewable fuels. **(2 marks)**

....

....

....

c) Name one renewable fuel. **(1 mark)**

....

The diagram below shows the main features of a gas power station.

d) Describe the energy change that takes place in the boiler. **(1 mark)**

....

....

....

e) Describe the energy change that takes place in the turbine section. **(1 mark)**

....

....

....

f) Describe the energy change that takes place in the generator section. **(1 mark)**

....

....

g) Why does the electrical energy produced by the generator pass through a transformer before going into the National Grid? **(3 marks)**

....

....

....

....

....

2 In the UK some of our electricity is generated by nuclear power stations.

a) Give one advantage of using nuclear power stations rather than fossil fuel power stations to generate electricity. **(1 mark)**

....

....

b) Give two disadvantages of using nuclear power stations rather than fossil fuel power stations to generate electricity. **(2 marks)**

....

....

....

....

How well did you do? ✗ 0-13 **Try again** 14-19 **Getting there** 20-26 **Good work** 27-33 **Excellent!** ✓

Renewable sources of energy

A

1 Which of the following is not a renewable source of energy? **(1 mark)**

a) tidal
b) wood
c) coal
d) waves

2 Which of these alternative sources of energy is not affected by the weather? **(1 mark)**

a) wind
b) solar
c) geothermal
d) biomass

3 Which of the following may be a concern for a group of people living on an isolated island who choose to use just solar cells to produce all their electricity? **(1 mark)**

a) noise pollution
b) no electricity available at night
c) visual pollution
d) poor energy capture

4 Which of the following is a source of energy obtained from 'things that were once alive'? **(1 mark)**

a) wave
b) hydroelectric
c) geothermal
d) biomass

5 The source of most of the energy on the Earth **(1 mark)**

a) the Moon
b) the Sun
c) the oceans
d) the soil

B

1 Draw a line from each of the seven alternative sources of energy a) to a box containing an advantage of using this source and b) to a box containing a disadvantage. **(14 marks)**

Advantage of using this source	Alternative source of energy	Disadvantage of using this source
using this fuel does not add to the greenhouse effect	GEOTHERMAL	obstacle to water traffic
only low level technology is needed	TIDAL	large area of land needed for renewal of supply
energy can be stored until needed	SOLAR	very high initial construction costs
useful for isolated island communities	BIOMASS	few suitable sites
reliable, available twice a day	WIND	poor energy capture therefore large area needed
no pollution	HYDROELECTRIC	possible visual and noise pollution
no pollution and no environmental problems	WAVE	not useful where there is limited sunshine

C

1 The diagram below shows a hydroelectric power station.

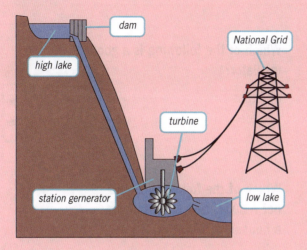

dam

National Grid

high lake

turbine

station gernerator

low lake

a) What kind of energy does the water stored in the top lake have? **(1 mark)**

....

....

....

b) What kind of energy does the water have after it has fallen and reaches the turbine?

(1 mark)

....

....

....

c) What happens to this energy at the station generator? **(1 mark)**

....

....

....

d) Name one disadvantage using this source of energy may have on the environment.

(1 mark)

....

....

e) Give one advantage of using this method of providing electrical energy. **(1 mark)**

....

....

f) When 2000 J of energy enter the station generator 1000 J of energy are produced for the National Grid. Calculate the efficiency of the energy transfer. **(2 marks)**

....

....

....

2 Burning fuels such as coal and oil increases the amount of carbon dioxide in the atmosphere and so increases the greenhouse effect.

a) What is the greenhouse effect? **(1 mark)**

....

....

b) How could we reduce the amount of carbon dioxide we are releasing into the atmosphere? **(1 mark)**

....

....

How well did you do? ✗ 0-11 **Try again** 12-17 **Getting there** 18-22 **Good work** 23-28 **Excellent!** ✓

Heat transfer – conduction

A

1 **Heat will flow:** *(1 mark)*

a) from hot places to cold places
b) from cold places to hot places
c) quickly through an insulator
d) slowly through a conductor

2 **A good example of a conductor is:** *(1 mark)*

a) plastic
b) wood
c) copper
d) paper

3 **The transfer of heat by conduction cannot take place through a** *(1 mark)*

a) solid
b) liquid
c) gas
d) vacuum

4 **Which of the following is a good insulator?** *(1 mark)*

a) brass
b) steel
c) air
d) gold

5 **Which of the following would not reduce the heat loss from a house?** *(1 mark)*

a) installing double glazing
b) turning up the thermostat on the central heating
c) putting insulation in the loft
d) fitting draught excluders

B

1 The diagram below shows how heat is lost from an un-insulated home. Unfortunately the words and numbers for all five labels have become mixed up. Using only the words given below, write out the correct labels in the correct positions around the house. Each word and number should be used only once. **(10 marks)**

reduced by	insulation	through floor	and cracks	25%	wall
reduced by	through walls	and underlay	through	15%	insulation
reduced by	having cavity	fitting draft	excluders	25%	
reduced by	and windows	through gaps	installing	putting	
reduced by	into loft	double glazing	10%	windows	
roof	fitting carpets	around doors	25%	through	

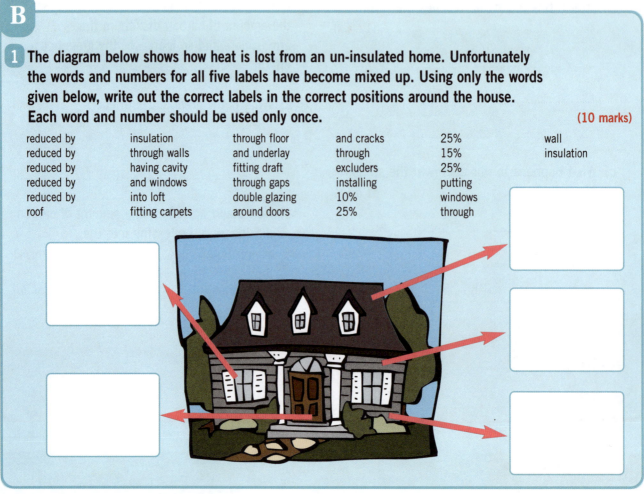

C

1 Match the words in this list with the descriptions given in the table below.   (5 marks)

conduction
insulator
air
fibreglass
aluminium

Description	Word
Used in places where you want to prevent heat flow	
The movement of heat through a solid by vibrations	
Lots of this is trapped in warm woollen clothing	
An example of a good conductor	
Can be used to prevent heat loss through the roof	

2 Richard wants to insulate his home. Before he does so, he searches the internet to find out how much each type of insulation will cost and how much each will save him annually. The table below shows the information he has obtained.

Type of insulation	Typical cost	Typical annual saving	Payback time
double glazing	£2500	£100	
loft insulation	£200	£100	
draft excluders	£100	£25	
cavity wall insulation	£500	£50	

a) Work out the payback time for each type of insulation.
   Write each of your answers in the correct space in the table.   (4 marks)

b) Which type of insulation is most cost effective? Explain your answer.   (2 marks)

....

....

....

....

....

Heat transfer – convection

A

1 The circular movement of air caused by heat is called: **(1 mark)**

a) transpiration

b) evaporation

c) radiation

d) a convection current

2 2 When air is warmed it: **(1 mark)**

a) expands and falls

b) contracts and rises

c) expands and rises

d) contracts and falls

3 In which kinds of materials can convection not take place? **(1 mark)**

a) gases

b) liquids

c) fluids

d) solids

4 To heat the whole of an oven the heater must be placed: **(1 mark)**

a) at the top

b) in the middle

c) at the bottom

d) at the back

5 Cavity wall insulation reduces heat loss from a house because: **(1 mark)**

a) it increases the amount of air between the walls

b) it stops the cold entering the house

c) it prevents convection currents between the walls

d) it reduces the amount of air between the walls

B

1 The incomplete diagrams to the right show how different sea breezes are created at the coast.

onshore breeze

offshore breeze

a) Fill in the missing words. **(6 marks)**

During the .... when the sun is shining, the land becomes warmer than the .... Air above the land is warmed, it .... and .... Cooler air moves from just above the .... to take the place of the rising air. This is why on a hot day we often feel an onshore .... at the seaside.

b) Explain how a convection current is set up at the seaside during the night. **(3 marks)**

....

....

....

C

1 **a)** Explain why convection currents can transfer heat in liquids and gases but not through solids. **(2 marks)**

....

....

....

....

b) The diagram below shows a cavity wall.

What happens to the air next to the inner wall? **(2 marks)**

....

....

....

....

c) What is set up between the walls? **(1 mark)**

....

....

....

....

d) Why is it difficult for heat to cross the gap by conduction? **(1 mark)**

....

....

....

....

e) What can be done to prevent heat loss across the gap by convection? Explain why this prevents heat loss. **(2 marks)**

....

....

....

2 The picture below is of an open fire.

a) Explain why open fires like the one above are not very efficient at warming a room. **(2 marks)**

....

....

....

....

....

....

Heat transfer – radiation

A

1 Radiation is the transfer of heat by **(1 mark)**

a) vibrating particles ☐
b) oscillating particles ☐
c) waves ☐
d) evaporation ☐

2 When radiation strikes a dark rough surface it is likely to be **(1 mark)**

a) absorbed ☐
b) reflected ☐
c) refracted ☐
d) diffracted ☐

3 When radiation strikes a light shiny surface it is likely to be **(1 mark)**

a) diffracted ☐
b) reflected ☐
c) refracted ☐
d) absorbed ☐

4 A thermogram is **(1 mark)**

a) a kind of heater ☐
b) an insulator ☐
c) a kind of photograph ☐
d) a measurement of heat ☐

5 To try to keep cool, people in hot countries could paint the outside of their houses **(1 mark)**

a) green ☐
b) black ☐
c) white ☐
d) brown ☐

B

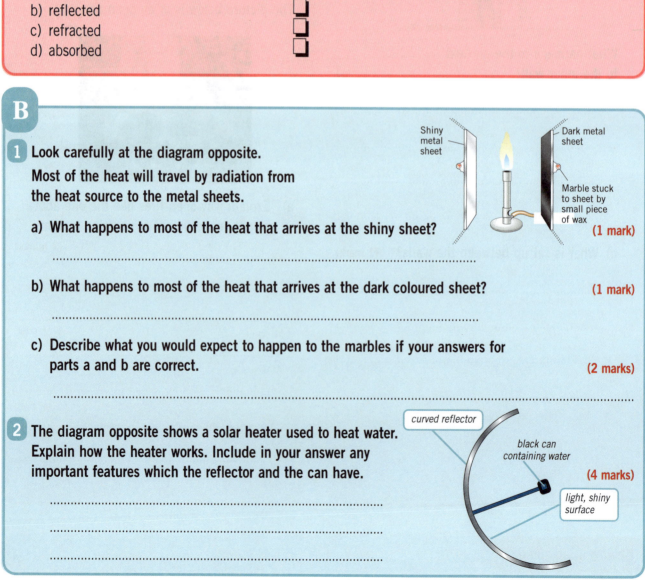

1 Look carefully at the diagram opposite.

Most of the heat will travel by radiation from the heat source to the metal sheets.

Shiny metal sheet

Dark metal sheet

Marble stuck to sheet by small piece of wax

a) What happens to most of the heat that arrives at the shiny sheet? **(1 mark)**

....

b) What happens to most of the heat that arrives at the dark coloured sheet? **(1 mark)**

....

c) Describe what you would expect to happen to the marbles if your answers for parts a and b are correct. **(2 marks)**

....

2 The diagram opposite shows a solar heater used to heat water. Explain how the heater works. Include in your answer any important features which the reflector and the can have. **(4 marks)**

curved reflector

black can containing water

light, shiny surface

....
....
....

C

1 The diagram below shows two cars which have been parked outside in the sun all day.

a) Which of the two cars will be hottest to touch? **(1 mark)**

....

....

....

b) Explain why the car you have chosen in part a) will be the hottest. **(2 marks)**

....

....

....

....

....

To prevent the interior of the car from becoming too hot John buys a screen made of card which he places inside the front window of his car.

c) What colour should the card be? **(1 mark)**

....

....

....

d) Explain why the card should be the colour you have chosen in answer c). **(2 marks)**

....

....

....

....

2 The diagram below shows the construction of a thermos flask.

plastic stopper

silvered surface

vacuum

Describe how each of the following parts reduces heat loss from the flask.

a) plastic stopper **(1 mark)**

....

....

b) vacuum **(1 mark)**

....

....

c) silvered surfaces **(1 mark)**

....

....

How well did you do? ✗ 0-9 **Try again** 10-13 **Getting there** 14-18 **Good work** 19-22 **Excellent!** ✓

Current, charge and resistance

A

1 An electric current is a flow of **(1 mark)**

 a) atoms
 b) charge
 c) water
 d) molecules

2 Which of the following is not true. The current we get from a battery **(1 mark)**

 a) is a.c.
 b) is d.c.
 c) flows from positive to negative
 d) flows in one direction only

3 If a battery is labelled 20 amp-hour this means **(1 mark)**

 a) it can provide a maximum current of 20 A
 b) it can provide a current for 20 hours
 c) it can provide a current of 4 A for 5 hours
 d) it can provide a current of 20 A for 20 hours

4 A resistor which could be used to adjust the brightness of the lights in a room is called **(1 mark)**

 a) a fixed resistor
 b) a thermistor
 c) a thermostat
 d) a variable resistor

5 A resistor which could be used to detect temperature changes is called **(1 mark)**

 a) a diode
 b) a thermomstat
 c) a thermistor
 d) a filament bulb

B

1 Calculate the resistance of each of the bulbs shown below. **(5 marks)**

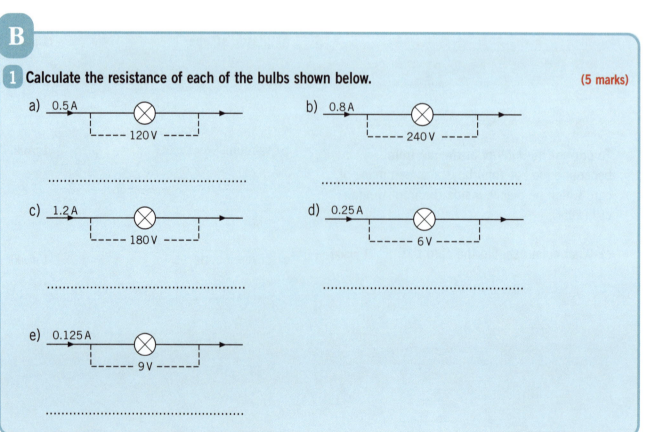

a) 0.5 A 120 V

b) 0.8 A 240 V

c) 1.2 A 180 V

d) 0.25 A 6 V

e) 0.125 A 9 V

....

C

1 The diagram below shows a circuit containing a battery, a light bulb and a variable resistor.

a) What is a variable resistor? (1 mark)

....

....

b) How can the brightness of the bulb be reduced? (1 mark)

....

....

c) Name a public building where variable resistors and used in this way. (1 mark)

....

....

d) When a p.d. of 12 V is applied across a variable resistor a current of 0.2 A passes through it. Calculate the resistance of the variable resistor. (2 marks)

....

....

....

....

2 The diagram below shows a circuit containing a battery, a buzzer and a light dependent resistor (LDR).

a) What is a light-dependent resistor? (1 mark)

....

....

b) Describe how a circuit like that shown above could be used as a simple burglar alarm. (3 marks)

....

....

....

....

....

....

c) Calculate resistance of the LDR if a current of 0.05 A flows when a potential difference of 6 V is applied across it. (2 marks)

....

....

....

....

Electrical power

A

1 The power rating of a light bulb tells us **(1 mark)**

a) the size of the light bulb ☐

b) the total energy changed into light by the bulb ☐

c) the efficiency of the bulb ☐

d) how much energy the bulb changes each second ☐

2 A 60 W electric bulb changes **(1 mark)**

a) 60 J of electrical energy into heat and light energy every second ☐

b) 60 C of electrical energy into heat and light energy every second ☐

c) 60 J of electrical energy into heat and light energy every hour ☐

d) 60 C of electrical energy into heat and light energy every minute ☐

3 Calculate the energy used in units when a 2 kW fire is turned on for six hours. **(1 mark)**

a) 3 units ☐

b) 12 units ☐

c) 4 units ☐

d) 12 000 units ☐

4 Calculate the cost of turning on a 3 kW tumble dryer for two hours, if the cost of one unit of electricity is 10p. **(1 mark)**

a) 6p ☐

b) 15p ☐

c) 60p ☐

d) 66p ☐

5 What device in your house tells you how much electrical energy you have used? **(1 mark)**

a) gas meter ☐

b) electricity meter ☐

c) ammeter ☐

d) fuse box ☐

B

1 The wordsearch below contains eight words or phrases connected to this topic. Can you find them all?

(8 marks)

P	S	T	A	N	D	I	N	G	M
O	C	H	A	R	G	E	H	I	E
W	U	K	I	L	O	W	A	T	T
E	N	H	O	U	R	G	K	J	E
R	I	C	E	F	B	I	L	L	R
B	T	D	E	N	E	R	G	Y	L
A	P	P	L	I	A	N	C	E	M

C

ELECTRICITY BILL

Charges for electricity used

Present reading 75889	Previous reading 74489	Units used	Pence per unit 10.00	Charge amount £
Quarterly standing charge				£ 15.00
Total				£

1 Julia has just received her electricity bill for the last quarter. The bill is shown above.

a) How many units of electrical energy has Julia used in this quarter? **(1 mark)**

....

....

b) What is the cost of this energy if 1 unit costs 10p? **(1 mark)**

....

....

c) Explain why Julia has to pay a £15.00 standing charge. **(1 mark)**

....

....

d) What is Julia's total bill? **(1 mark)**

....

....

e) How many times each year will Julia receive a bill like this? **(1 mark)**

....

....

Julia turned on a 2 kW heater at eight o'clock in the morning. She forgot to turn it off before she went to work. She returned home at six o'clock in the evening.

f) How long was the heater turned on? **(1 mark)**

....

....

....

....

g) How may units of electrical energy did the heater use during this time? **(1 mark)**

....

....

....

....

h) What was the cost of this energy if 1 unit costs 10p? **(1 mark)**

....

....

....

....

How well did you do? ✗ **0-8** Try again   **9-13** Getting there   **14-17** Good work   **18-21** Excellent! ✓

Electric motors

A

1 Which of the following does not contain an electric motor? **(1 mark)**

- a) cassette player ☐
- b) DVD player ☐
- c) radio ☐
- d) food mixer ☐

2 Which three ingredients are absolutely necessary for an electric motor? **(1 mark)**

- a) current, resistance and wire ☐
- b) coil, current and resistance ☐
- c) coil, current and magnetic field ☐
- d) current, resistance and magnetic field ☐

3 Which of the following statements is true for a current carrying wire placed between the poles of a magnet? **(1 mark)**

- a) The wire will move from a strong part of the magnetic field to a weaker part. ☐
- b) The wire may move from the north pole of the magnet to the south pole. ☐
- c) The wire may move from the south pole of the magnet to the north pole. ☐
- d) The wire will be pushed into the strongest part of the field between the poles of the magnet. ☐

4 A loop of wire placed between the poles of a magnet rotates when current passes through it because the forces on opposite sides of the loop **(1 mark)**

- a) are equal in strength and direction ☐
- b) are both upward ☐
- c) are both downward ☐
- d) are in opposite directions ☐

5 Which of the following will not make a motor turn more quickly? **(1 mark)**

- a) Increase the current ☐
- b) Increase the strength of the magnet ☐
- c) Increase the number of turns on the coil ☐
- d) Replace the split ring with two full rings ☐

B

1 The diagrams below show currents passing through wires lying between the poles of a magnet.

(a)   N   S   force
(b)   N   S
(c)   S   N
(d)   S   N

a) Each of the wires will experience a force. The direction of the force on the first wire is shown. Determine the direction of the force for each wire.

i) .... **(1 mark)**

ii) .... **(1 mark)**

iii) .... **(1 mark)**

b) State two ways in which the size of this force could be increased. **(2 marks)**

....

C

1 A loop of wire is placed between the poles of a magnet.

a) Explain why the loop begins to rotate when current flows through it. **(3 marks)**

....

....

....

....

....

b) Why does the loop stop rotating when it is vertical? **(2 marks)**

....

....

....

....

....

....

c) Explain how the use of a split ring allows the coil to rotate continuously. **(3 marks)**

....

....

....

....

....

....

d) Suggest three ways in which the speed of a motor can be increased. **(3 marks)**

....

....

....

....

e) Describe two ways in which real motors are different to the simple motor described above. **(2 marks)**

....

....

....

....

f) What would happen to a motor if the connections to the power supply were reversed? **(1 mark)**

....

....

....

....

Generators and alternators

A

1 Which of the following is not used to generate a current? **(1 mark)**

a) generator ❑
b) electric motor ❑
c) alternator ❑
d) dynamo ❑

2 Which of the following will produce a voltage across a wire? **(1 mark)**

a) moving the wire from the south pole of a magnet to the north ❑
b) moving the wire from the north pole of a magnet to the south pole ❑
c) holding the wire stationary between the poles of a magnet ❑
d) moving the wire across magnetic field lines ❑

3 Which of the following statements is untrue for an alternator? **(1 mark)**

a) An alternator produces induced currents. ❑
b) An alternator produces alternating current. ❑
c) An alternator produces current which is continually changing direction. ❑
d) An alternator produces direct current. ❑

4 When a cyclist stops, his dynamo will produce no current because: **(1 mark)**

a) the circuit is incomplete ❑
b) there is no need for electricity until he starts to move ❑
c) there is no movement of the magnet ❑
d) his lights are no longer connected to the dynamo ❑

5 Which of the following will not increase the current produced by a generator? **(1 mark)**

a) using a stronger magnet ❑
b) using thicker brushes ❑
c) turning the coil more quickly ❑
d) using a coil with more turns ❑

B

1 How many of these anagrams can you solve? **(10 marks)**

lcio ....

ymnoda ....

govtlae ....

galtrnnatie ....

rreuntc ....

gcitnaem eidfl ....

oaaeltrntr ....

ytteelcriic ....

eeongratr ....

neicdu ....

2 The diagram opposite shows a simple dynamo.

a) Give one use for a simple dynamo like this. **(1 mark)**

....

b) Explain what happens when the wheel stops turning. **(2 marks)**

....

....

knurled knob

soft iron core

rotating magnet

coil

C

1 The diagram below shows a magnet being pushed into a long coil. As the magnet moves into the coil, the needle of the galvanometer is seen to move a little to the right.

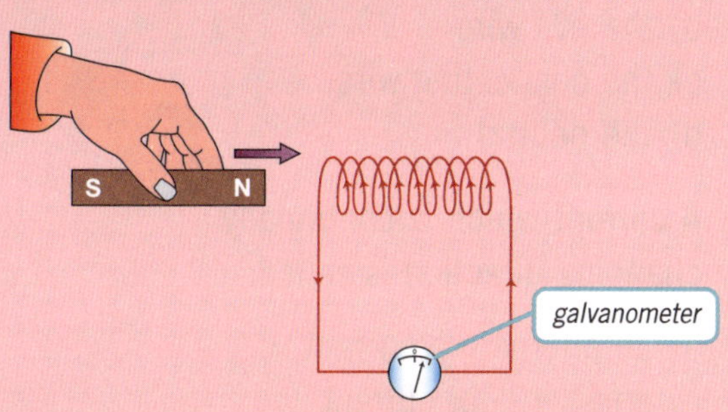

galvanometer

Describe what would happen if

a) The magnet is withdrawn from the coil quickly. (2 marks)

b) The south pole of the magnet is pushed into the coil slowly. (2 marks)

c) The magnet is held stationary inside the coil. (1 mark)

d) The magnet is held stationary but the coil is moved to the right. (1 mark)

2 a) What is alternating current? (1 mark)

The diagram below shows a simple alternator.

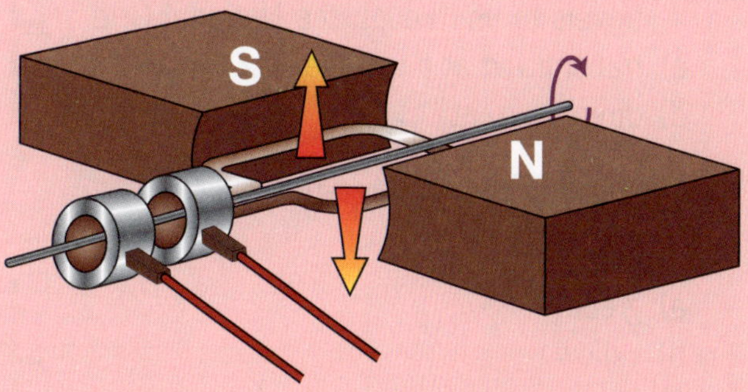

b) Explain in detail how the alternator produces an alternating current. (4 marks)

c) State three ways in which the current being produced by the alternator could be increased. (3 marks)

Domestic electricity

A

1 The mains electricity we use in the home is: (1 mark)

a) identical to the electricity we get from cells and batteries ☐

b) generated at a power station and then travels to us through the National Grid ☐

c) transmitted at a voltage of 24 V ☐

d) direct current ☐

2 The earth wire in a three-pin plug is: (1 mark)

a) green ☐

b) yellow ☐

c) brown ☐

d) green and yellow ☐

3 The wire through which electrical energy travels to an appliance is known as: (1 mark)

a) the earth wire ☐

b) the live wire ☐

c) the connecting wire ☐

d) the neutral ☐

4 A current which is continually changing direction is called: (1 mark)

a) an alternating current ☐

b) an alternator current ☐

c) a direct current ☐

d) an induced current ☐

5 All three pin plugs in the UK contain: (1 mark)

a) a resistor ☐

b) a capacitor ☐

c) a fuse ☐

d) an inductor ☐

B

1 The diagram opposite shows the inside of a three-pin plug.

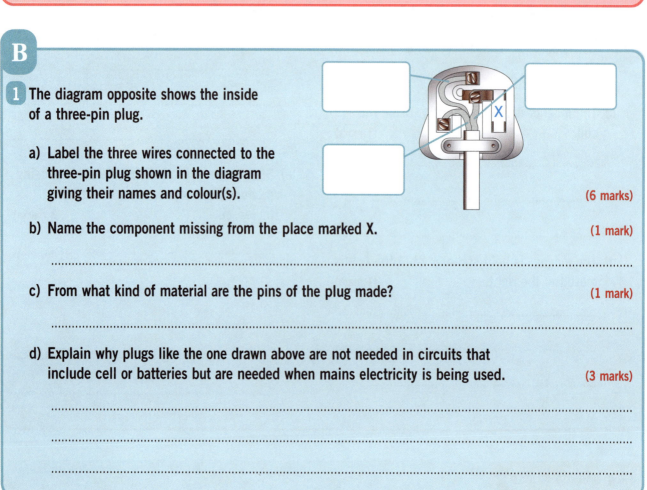

a) Label the three wires connected to the three-pin plug shown in the diagram giving their names and colour(s). (6 marks)

b) Name the component missing from the place marked X. (1 mark)

....

c) From what kind of material are the pins of the plug made? (1 mark)

....

d) Explain why plugs like the one drawn above are not needed in circuits that include cell or batteries but are needed when mains electricity is being used. (3 marks)

....

....

....

C

1 The diagram below shows a kettle which has a metal outer casing.

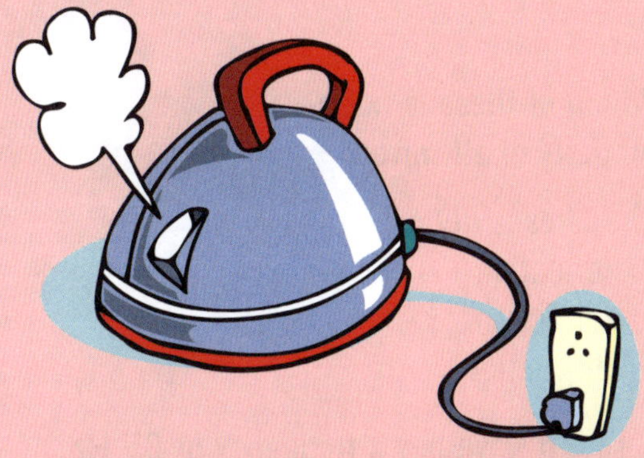

a) Explain what might happen to someone using this type of kettle if the heating element is faulty and there is no earth wire connected. (1 mark)

....

....

b) What will happen if an earth wire is connected to the metal casing? (1 mark)

....

....

c) What is the name of the kind of insulation provided by a kettle which has a plastic casing? (1 mark)

....

....

d) Why do circuits contain fuses? (3 marks)

....

....

....

....

e) Explain how a fuse works. ....(2 marks)

....

....

....

....

....

f) Give the values of two fuses commonly used in the UK. (2 marks)

....

....

g) What does a circuit breaker do? (1 mark)

....

....

h) Give one example of when someone might use a circuit breaker. (1 mark)

....

....

2 a) What is direct current? (1 mark)

....

b) What is alternating current? (1 mark)

....

c) Give one example of a source of direct current. (1 mark)

....

d) Give one example of a source of alternating current. (1 mark)

....

How well did you do? ✗ 0-12 Try again 13-19 Getting there 20-25 Good work 26-32 Excellent! ✓

Waves

A

1 What do waves carry from place to place? *(1 mark)*

a) sound ☐
b) light ☐
c) energy ☐
d) electricity ☐

2 Which of the following are not transverse waves? *(1 mark)*

a) sound waves ☐
b) X-rays ☐
c) radio waves ☐
d) microwaves ☐

3 The frequency of a wave is: *(1 mark)*

a) the height of the wave ☐
b) the distance between the peak of one wave and the peak of the next ☐
c) the number of waves made by the source each second ☐
d) the distance between a peak and a trough ☐

4 Which of these is not a property of all waves? *(1 mark)*

a) radioactivity ☐
b) reflection ☐
c) refraction ☐
d) diffraction ☐

5 A sound wave has a frequency of 85 Hz and a wavelength of 4 m. Calculate the velocity of this wave. *(1 mark)*

a) 720 m/s ☐
b) 21 m/s ☐
c) 330 m/s ☐
d) 340 m/s ☐

B

1 There are 13 words connected to this topic hidden in this wordsearch. Can you find them all? *(13 marks)*

R	E	F	L	E	C	T	E	D	E	L	S
E	V	I	B	R	A	T	I	O	N	I	P
F	R	E	Q	U	E	N	C	Y	A	G	S
R	W	A	V	E	L	E	N	G	T	H	O
A	H	S	E	I	S	M	I	C	S	T	U
C	E	A	B	R	I	A	N	T	P	B	N
T	R	A	N	S	V	E	R	S	E	M	D
E	T	J	I	L	L	Y	N	T	E	R	T
D	Z	A	M	P	L	I	T	U	D	E	Z
L	O	N	G	I	T	U	D	I	N	A	L

C

1 The diagram below shows a transverse wave.

A transverse wave

a) Mark accurately on the diagram
   i) the wavelength of the wave and
   ii) the amplitude of the wave.   **(2 marks)**

b) What is a transverse wave?   **(1 mark)**

....

....

c) Give one example of a transverse wave.
        (1 mark)

....

....

....

d) What is a longitudinal wave?   **(1 mark)**

....

....

....

e) Give one example of a longitudinal wave.
        (1 mark)

....

....

....

f) The wave drawn above has a frequency of
   25 Hz and a wavelength of 8 m.
   Calculate the speed of this wave.   **(3 marks)**

....

....

....

....

....

2 a) Name one source of seismic waves.   **(1 mark)**

....

b) Name two types of seismic waves.   **(2 marks)**

....

....

....

....

c) Describe one difference between the two
   waves you have named in part b).   **(2 marks)**

....

....

....

....

d) How have these waves been useful to
   scientists?   **(1 mark)**

....

....

....

How well did you do?   ✗ 0-13 Try again   14-19 Getting there   20-26 Good work   27-33 Excellent! ✓

Electromagnetic spectrum 1

A

1 Which of the following types of waves has the longest wavelength? (1 mark)

a) gamma rays ☐
b) X-rays ☐
c) radio waves ☐
d) microwaves ☐

2 Which of the following types of waves has the highest frequency? (1 mark)

a) radio waves ☐
b) visible light ☐
c) infrared waves ☐
d) gamma rays ☐

3 Which of the following statements is untrue? (1 mark)

a) All members of the electromagnetic spectrum travel at the same speed through a vacuum. ☐
b) All members of the electromagnetic spectrum transfer energy. ☐
c) All members of the electromagnetic spectrum affect living cells. ☐
d) All members of the electromagnetic spectrum can be reflected. ☐

4 The properties of the different groups of the electromagnetic spectrum vary because (1 mark)

a) they have different wavelengths ☐
b) they have different amplitudes ☐
c) they travel at different speeds ☐
d) they are absorbed by different amounts ☐

5 Which type of waves are used by mobile phones? (1 mark)

a) ultraviolet ☐
b) infrared ☐
c) microwaves ☐
d) gamma rays ☐

B

1 a) Unscramble the letters to reveal seven parts of the electromagnetic spectrum. (7 marks)

vsrdweoaai ....

rvmcweaios ....

lraioelttvu ....

eaindfrr ....

axsyr ....

smmaaaygr ....

gveiiistbllh ....

b) Unscramble the letters to reveal properties that are common to all the different waves that make up the electromagnetic spectrum. (5 marks)

rntsearsev ....

erractinof ....

mntaisssrion ....

eeflctinro ....

asnoobrpti ....

C

1 The diagram below shows the electromagnetic spectrum.

radio waves   microwaves   infrared   ultraviolet   gamma rays

A   B

a) What are the names of the two groups of waves A and B? (2 marks)

....

....

....

....

b) Which group of waves has no known effect on living cells? (1 mark)

....

....

....

c) Which group of waves do we use to see? (1 mark)

....

....

....

d) Which two groups of waves are the most penetrating? (2 marks)

....

....

....

....

2 Explain in your own words what happens to a wave when it strikes an object and is

a) transmitted (1 mark)

....

....

b) reflected (1 mark)

....

....

c) absorbed (1 mark)

....

....

You can draw diagrams to illustrate your answers if you wish.

d) Infrared radiation is shone onto an object. Name two factors which will decide the amount of radiation the object will absorb. (2 marks)

....

....

....

e) Jake sits in his garden sunbathing. Describe three effects the ultraviolet radiation from the Sun may have on his skin. (3 marks)

....

....

....

....

How well did you do?   ✗ 0-12 Try again   13-19 Getting there   20-25 Good work   26-31 Excellent! ✓

Electromagnetic spectrum 2

A

1 Which of the following types of waves are used for cooking? **(1 mark)**

a) gamma rays ☐
b) X-rays ☐
c) ultraviolet waves ☐
d) microwaves ☐

2 Which of the following waves do we use to see? **(1 mark)**

a) radio waves ☐
b) visible light ☐
c) infrared waves ☐
d) microwaves ☐

3 Which group of waves is used to look inside a body? **(1 mark)**

a) ultraviolet waves ☐
b) X-rays ☐
c) microwaves ☐
d) radio waves ☐

4 Which of the following is not a use of infrared waves? **(1 mark)**

a) sun lamps ☐
b) remote controls for TV ☐
c) night vision cameras ☐
d) heating lamps ☐

5 Which of the following is not a use of ultraviolet waves? **(1 mark)**

a) sun lamps ☐
b) detecting forged bank notes ☐
c) night vision cameras ☐
d) checking for security markings ☐

B

1 Which of these statements are true and which are false? **(10 marks)**

	true	false
a) Gamma rays are used to communicate over large distances.	☐	☐
b) Heat seeking cameras create images using infrared waves.	☐	☐
c) Microwaves cause some chemicals to fluoresce.	☐	☐
d) Radio waves can carry information in two forms, analogue and dialogue.	☐	☐
e) Microwaves cook food by first warming the outside.	☐	☐
f) Modern telecommunication networks use optical fibres to carry messages.	☐	☐
g) Ultrasonic scanning is now used to monitor the development of a foetus during pregnancy as it is less dangerous than using X-rays.	☐	☐
h) Gamma rays are used for radiotherapy.	☐	☐
i) Forged bank notes can be detected using infrared waves.	☐	☐
j) X-rays have a very short wavelength and a very high frequency.	☐	☐

C

1 The diagram below shows the positions of a radio transmitter and receiver.

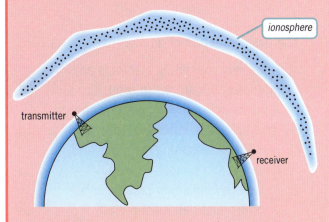

a) What is the ionosphere? **(1 mark)**

....

b) Draw lines to the diagram above to show how we can use the ionosphere to send long wavelength radio messages over large distances. **(3 marks)**

....

....

....

c) Draw a diagram in the space below to show how microwaves can be used to send signals over large distances. **(3 marks)**

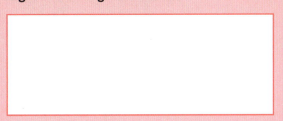

d) Why are most modern radio signals sent in digital form rather than in analogue?

(1 mark)

....

....

2 Bill changes the channel on his television using his remote control.

a) What kind of waves is emitted by Bill's remote when he presses a button? **(1 mark)**

....

b) Bill normally points the remote at the television when he wants to change channel but he has discovered that he can also change channels by pointing the remote away from the television and towards a part of the wall which is behind him. What wave property is Bill using to get his signal from the remote to the television? **(1 mark)**

....

Bill goes into the next room. He points his remote directly at the television in the other room.

 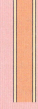 

c) Explain why Bill is unable to change the television channel using his remote.

(2 marks)

....

....

....

How well did you do? ✗ **0-11** Try again   **12-16** Getting there   **17-22** Good work   **23-27** Excellent! ✓

Nuclear radiation

A

1 Which of the following statements is true for alpha particles? **(1 mark)**

a) They travel at the speed of light. ❑
b) They are unaffected by electric fields. ❑
c) They are poor ionisers. ❑
d) They are helium nuclei. ❑

2 Which of the following statements is true for beta particles? **(1 mark)**

a) They are fast moving electrons. ❑
b) They are uncharged. ❑
c) They are more penetrating than gamma rays. ❑
d) They are unaffected by magnetic fields. ❑

3 Which of the following statements is true for gamma rays? **(1 mark)**

a) They are deflected by magnetic fields. ❑
b) They are not very penetrating. ❑
c) They travel at the speed of light. ❑
d) They are deflected by electric fields. ❑

4 The basic structure of an atom is: **(1 mark)**

a) A nucleus containing electrons and protons and neutrons in orbit. ❑
b) A nucleus containing neutrons and protons and electrons in orbit. ❑
c) A nucleus containing electrons and neutrons and protons in orbit. ❑
d) A nucleus containing protons and electrons and neutrons in orbit. ❑

5 Which of the following statements is untrue? **(1 mark)**

a) Alpha radiation is the least penetrating radiation. ❑
b) Beta radiation is the most penetrating radiation. ❑
c) Gamma radiation is more penetrating than alpha radiation. ❑
d) Beta radiation is more penetrating than alpha radiation. ❑

B

1 Which of the following statements are true and which are false? **(10 marks)**

	true	false
a) Alpha particles are fast moving, high penetrating helium nuclei.	❑	❑
b) Gamma waves move at the speed of light.	❑	❑
c) Beta particles are not deviated by magnetic and electric fields.	❑	❑
d) Alpha particles are not deviated a lot by magnetic and electric fields because they are moving very quickly.	❑	❑
e) Gamma radiation produces lots of ions as it passes through objects.	❑	❑
f) Gamma radiation is uncharged and therefore not affected by magnetic or electric fields.	❑	❑
g) Alpha radiation is very similar to X-rays.	❑	❑
h) Beta particles are more penetrating than alpha particles because they are smaller.	❑	❑
i) Alpha radiation is unable to travel more than a few centimetres in air.	❑	❑
j) An ion is an atom that has equal numbers of protons and electrons.	❑	❑

C

1 a) Which part of an atom emits radiation? (1 mark)

The diagram below shows a radioisotope which is emitting two different types of radiation.

thin sheet of aluminium
A
B
thick sheets of lead
card

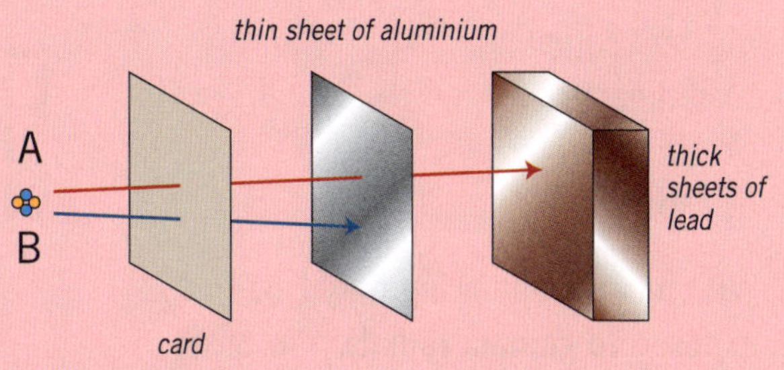

b) Identify on the diagram the different types of radiation being emitted. (2 marks)

c) What is an ion? (1 mark)

d) Explain how radiation creates ions. (2 marks)

e) Write down the names of the three different types of radiation in the order of their ionising power, i.e. the best ioniser first and the poorest last. (1 mark)

2 The diagram below shows the effect of a magnetic field on all three types of radiation. Identify which path is followed by which radiation. (3 marks)

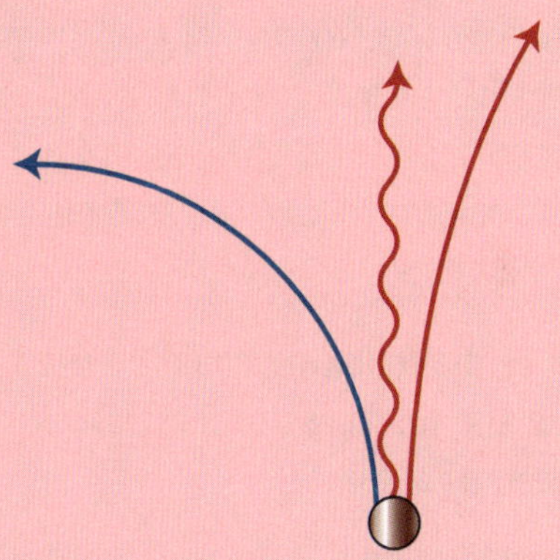

3 A helium atom contains two protons, two neutrons and two electrons. In the space below draw a picture of a helium atom. (3 marks)

How well did you do? 0-11 Try again 12-16 Getting there 17-22 Good work 23-28 Excellent!

Uses of radioactivity

A

1 In which of the following devices will you find a radioisotope emitting alpha particles? **(1 mark)**

a) microwave oven
b) smoke detector
c) refrigerator
d) mobile phone

2 In which of these situations would a radioactive tracer not be used? **(1 mark)**

a) To check the progress of food through the digestive system.
b) To check the flow of blood in the body.
c) To check for leaks in a gas pipe.
d) To kill germs.

3 Which of the following could be used to monitor the thickness of a metal produced as a sheet? **(1 mark)**

a) beta radiation
b) alpha radiation
c) gamma radiation
d) alpha or beta radiation

4 A narrow beam of radiation can be used to kill cancerous cells. This treatment is called **(1 mark)**

a) sterilisation
b) radioactivity
c) radiotherapy
d) radioscopy

5 Which of the following might be exposed to gamma radiation in order to kill any germs present? **(1 mark)**

a) surgical instruments
b) surgeon's hands
c) disinfectant
d) antiseptic

B

1 The diagram on the right shows the apparatus used to control the thickness of sheet metal. A description of how the apparatus works is given below but parts of it have been omitted. Fill in the missing sentences.

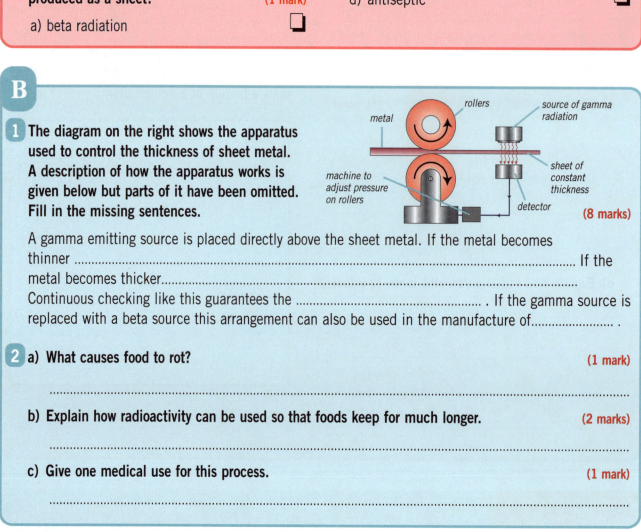

A gamma emitting source is placed directly above the sheet metal. If the metal becomes thinner .... If the metal becomes thicker...

Continuous checking like this guarantees the .... If the gamma source is replaced with a beta source this arrangement can also be used in the manufacture of........................ . **(8 marks)**

2 a) What causes food to rot? **(1 mark)**

....

b) Explain how radioactivity can be used so that foods keep for much longer. **(2 marks)**

....

c) Give one medical use for this process. **(1 mark)**

....

C

1 The diagram below shows how a radioactive source is used in radiotherapy.

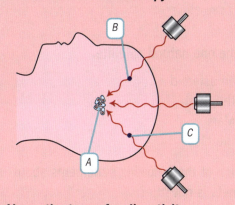

a) Name the type of radioactivity emitted by the source. **(1 mark)**

....

....

....

b) What happens to the cancerous cells at A? **(2 marks)**

....

....

....

....

c) Why does this not happen to cells at B or C? **(1 mark)**

....

....

....

2 The diagram below shows a man using a radiation detector. The liquid in the pipe has been labelled with a gamma emitter.

a) Suggest one material which might be flowing through the pipe. **(1 mark)**

....

b) What is the man trying to find? **(1 mark)**

....

....

c) How will he know when he finds it? **(1 mark)**

....

....

....

d) Why is the liquid not labelled with an alpha or beta emitter? **(1 mark)**

....

....

e) Give one advantage of this technique. **(2 marks)**

....

....

....

The Earth and our solar system

A

1 Which of the following contains most of the mass of our Solar System? (1 mark)

a) the Moon
b) the Sun
c) the Earth
d) the asteroids

2 Which of the following objects is luminous? (1 mark)

a) the Moon
b) the Sun
c) the Earth
d) comets

3 The planets are held in orbit around the Sun by (1 mark)

a) gravitational forces
b) electrostatic forces
c) magnetic forces
d) astronomical forces

4 Name one natural satellite. (1 mark)

a) an asteroid
b) the Earth
c) Mars
d) the Moon

5 Which of the following statements about comets is not true? (1 mark)

a) They travel fastest when furthest from the Sun.
b) They are made of rock-like pieces of ice.
c) A comet's tail is produced by ice that has melted.
d) They have very elliptical orbits.

B

1 There are 14 words or phrases connected to this topic hidden in this word search. Can you find them all? (14 marks)

S	A	T	U	R	N	M	J	A	L	N
A	O	C	Q	C	U	B	V	Q	R	Z
T	P	L	U	T	O	R	I	H	U	P
E	S	T	A	R	J	M	A	N	W	A
L	D	M	A	R	S	I	E	N	S	N
L	R	O	Q	U	S	K	H	T	U	S
I	F	O	V	G	L	Y	E	S	D	S
T	S	N	K	T	C	R	S	G	T	O
E	A	R	T	H	O	B	A	T	X	F
O	J	U	P	I	T	E	R	Y	E	E
M	E	P	D	W	O	R	B	I	T	M

C

1 The diagram above shows part of our solar system.

a) What is a comet? (2 marks)

....

....

b) What is an asteroid? (2 marks)

....

....

c) Where are most asteroids found? (1 mark)

....

....

d) How does the orbit of a comet differ from the orbit of a planet? (1 mark)

....

....

....

....

e) What kinds of forces keep objects in orbit around the Sun? (1 mark)

....

f) On the diagram above draw in the orbit of a comet. (1 mark)

g) Where in the orbit of a comet is it travelling fastest? (1 mark)

....

....

2 a) Explain the difference between a natural satellite and an artificial satellite. (1 mark)

....

....

b) Give two examples of the uses of artificial satellites. (2 marks)

....

....

....

How well did you do? ✗ 0-12 Try again 13-19 Getting there 20-25 Good work 26-31 Excellent! ✓

Stars and the universe

A

1 Our nearest star is: **(1 mark)**

a) the Sun ☐
b) the Moon ☐
c) the Milky Way ☐
d) Neptune ☐

2 A very large group of stars is called: **(1 mark)**

a) a nebula ☐
b) a constellation ☐
c) a solar system ☐
d) a galaxy ☐

3 Scientists believe that our universe began with a large explosion. They call this theory: **(1 mark)**

a) the Big Crunch ☐
b) the Big Explosion ☐
c) the Constantly Expanding Universe ☐
d) the Big Bang ☐

4 Which of the following is the correct sequence for the life of a star? **(1 mark)**

a) main stable period, nebula, red giant, white dwarf, black dwarf ☐
b) nebula, main stable period, red giant, white dwarf, black dwarf ☐
c) nebula, red giant, main stable period, black dwarf, white dwarf ☐
d) nebula, red giant, main stable period, white dwarf, black dwarf ☐

5 When a star forms, the energy it releases comes from: **(1 mark)**

a) nuclear reactions ☐
b) chemical reactions ☐
c) ionic reactions ☐
d) electronic reactions ☐

B

1 The four diagrams below show how stars and solar systems are formed. Write a brief explanation next to each diagram describing the processes that are taking place.

a) Stars begin to form when clouds of **(3 marks)**

....

....

....

b) This increase in.. **(2 marks)**

....

c) These reactions release .... **(2 marks)**

....

....

d) Smaller concentrations of................................... **(2 marks)**

....

....

C

1 Stars gradually change with time.

a) What forces pull matter together when a star first forms? **(1 mark)**

....

....

b) What kinds of reactions begin as the matter is pulled together? **(1 mark)**

....

....

c) What is released by these reactions? **(1 mark)**

....

....

d) How do we describe a star after it has been formed and the forces of expansion and the forces of attraction are balanced? **(1 mark)**

....

....

....

e) What is a supernova? **(1 mark)**

....

....

....

f) What remains at the end of the life of a star which has 'gone supernova'? **(1 mark)**

....

....

2 a) What is the Big Bang Theory? **(2 marks)**

....

....

....

....

....

b) What observations have scientists made that supports the Big Bang Theory? **(2 marks)**

....

....

....

....

....

c) Suggest two possible futures for the Universe. **(2 marks)**

....

....

....

....

3 a) What is a galaxy? **(1 mark)**

....

....

b) What is the name of the galaxy in which we live? **(1 mark)**

....

....

How well did you do? ✗ **0-11** Try again   **12-16** Getting there   **17-22** Good work   **23-28** Excellent! ✓

Exploring space

A

1 What is a flyby? (1 mark)

a) a kind of aircraft
b) a kind of probe
c) a kind of telescope
d) an asteroid

2 Which of the following is unlikely to be analysed using a lander? (1 mark)

a) soil samples
b) magnetic field strength
c) chemical content of atmosphere
d) speed of light in a vacuum

3 Large optical telescopes are often built on mountain tops because: (1 mark)

a) it is then closer to the stars and planets
b) it is quieter
c) there is less light pollution
d) there are no vibrations from passing traffic

4 Which of the following is not a benefit of exploring space? (1 mark)

a) smoke detectors
b) flat panel TVs
c) unleaded petrol
d) ultrasound scanners

5 Which of the following is the name of a telescope which is orbiting the Earth? (1 mark)

a) Big Brother
b) Sky satellite
c) Bubble telescope
d) Hubble telescope

B

1 Which of the following statements are true and which are false? (15 marks)

true false

a) Apart from the Earth, humans have only visited the Moon and Mars.
b) Large telescopes are built close to the coast where the air is cleaner.
c) Flyby probes don't land on a planet, moon, etc.
d) Landers are manned probes that actually land on moons and planets
e) Manned probes are cheaper than unmanned probes as there is less need for remote controls.
f) In cars, navigation systems are available because of space exploration.
g) Being weightless in space is a benefit to astronauts as they save energy.
h) Extra fuel has to be taken on manned missions compared with unmanned missions.
i) Muscle wastage in space can be avoided by following a balanced diet.
j) A lot of energy is needed to maintain a constant cabin temperature on manned missions.
k) Unmanned probes always have to return to Earth with all the information they have collected.
l) Radiation shields need to be included in the design of a manned probe.
m) We can see the moons of Jupiter with the naked eye.
n) Putting telescopes in orbit around the Earth allows us to see deeper into space.
o) Radio waves can be used to gather information about objects in space.

C

1 a) Explain why many telescopes are built on mountain tops. (1 mark)

....

....

....

b) Name three parts of the electromagnetic spectrum used by telescopes to gather information. (3 marks)

....

....

....

2 Probes are used to explore our solar system.

The photo below shows the probe Mars Observer which was sent to Mars in 1992.

a) Explain the difference between a flyby probe and a lander. (2 marks)

....

....

....

....

b) What kind of information might be gathered by a lander? (1 mark)

....

....

c) Why is most of the exploration of our solar system carried out by unmanned spacecraft and not manned spacecraft? (1 mark)

....

....

....

d) Name two problems that might be experienced by astronauts who have to live in zero gravity for more than just a few days. (2 marks)

....

....

....

....

e) Describe how astronauts overcome the problems you have described in part d). (1 mark)

....

....

f) Describe two other problems that have to be taken into account when planning a manned flight. (2 marks)

....

....

....

Notes

Notes